CHEMICAL REACTIONS OF
NATURAL AND SYNTHETIC POLYMERS

Ellis Horwood Series in
POLYMER SCIENCE AND TECHNOLOGY
Series Editors: T. J. KEMP, Professor of Chemistry, University of Warwick
J. F. KENNEDY, Professor of Chemistry, University of Birmingham

This series, which covers both natural and synthetic macromolecules, reflects knowledge and experience from research, development and manufacture within both industry and academia. It deals with the general characterisation and properties of materials from chemical and engineering viewpoints and will include monographs highlighting polymers of wide economic and industrial significance as well as of particular fields of application.

CHEMICAL REACTIONS OF NATURAL AND SYNTHETIC POLYMERS

M. LAZÁR, Ph.D., D.Sc.
T. BLEHA, Ph.D.
J. RYCHLÝ, Ph.D.
all of the Polymer Institute, Centre of Chemical Research
Slovak Academy of Sciences, Bratislava, Czechoslovakia

Translation Editor:
T. J. KEMP
Department of Chemistry, University of Warwick

ELLIS HORWOOD LIMITED
Publishers • Chichester

Halsted Press: a division of
JOHN WILEY & SONS
New York • Chichester • Brisbane • Toronto

ALFA
Publishers of Technical and Economic Literature • Bratislava

This English Edition first published in 1989 in coedition between
ELLIS HORWOOD LIMITED
Market Cross House, Cooper Street, Chichester, West Sussex, PO19 1EB, England
and
ALFA — Publishers of Technical and Economic Literature
Bratislava, Czechoslovakia

Distributors:

Australia and New Zealand:
JACARANDA WILEY LIMITED
GPO Box 859, Brisbane, Queensland 4001, Australia

Canada:
JOHN WILEY & SONS CANADA LIMITED
22 Worcester Road, Rexdale, Ontario, Canada

Europe and Africa:
JOHN WILEY & SONS LIMITED
Baffins Lane, Chichester, West Sussex, England

North and South America and the rest of the world:
Halsted Press: a division of
JOHN WILEY & SONS
605 Third Avenue, New York, NY 10158, USA

South-East Asia
JOHN WILEY & SONS (SEA) PTE LIMITED
37 Jalan Pemimpin # 05-04
Block B, Union Industrial Building, Singapore 2057

Indian Subcontinent
WILEY EASTERN LIMITED
4835/24 Ansari Road
Daryaganj, New Delhi 110002, India

© **1989 ALFA/Ellis Horwood Limited**

British Library Cataloguing in Publication Data
Lazár, M. (Milan), *1927—*
Chemical reactions of natural and synthetic polymers
1. Polymers. Functional groups. Chemical reactions.
I. Title   II. Bleha, T. (Thomas), *1943—*
III. Rychlý, J. (Josef), *1944—*
IV. Chemické reakcie. *English*

Library of Congress Card No. 88—13424

ISBN 0—7458—0193—5 (Ellis Horwood Limited)
ISBN 0—470—21231—4 (Halsted Press)

CONTENTS

5

6

PREFACE

The systematic investigation of chemical reactions of macromolecules started by the successful empirical modification of natural materials such as leather, rubber and cellulose. Further development of macromolecular chemistry as a science followed the synthesis of various polymers which either replaced or complemented the existing spectrum of construction and protective materials.

Research interest nowadays is focused on the preparation of polymers with tailor-made properties; for such a purpose the chemical modification of properties of pre-existing macromolecular systems is sometimes the most effective way. Particular attention is thus paid to biopolymers in their biological functions and in modelling of artificial bioanalogous systems.

In textbooks of macromolecular chemistry, the chemical reactions of macro-molecules are often eclipsed by the direct synthesis of polymers from low-mole-cular-mass monomers. Such a situation which has its rational and historical reasons prompted us to write a book devoted to chemical transformations of polymers. An attempt has been made to combine an introduction to the main principles of chemical reactions of macromolecules with an outlook to prospective developments of macromolecular chemistry as regards the enabling of more efficient technologies.

Our aim could not have been realized without the assistance of the staff of Publishing House Alfa, careful work of both reviewers prof. K. Veselý and dr. Z. Maňásek and without the support of colleagues from the Polymer Institute of Centre of Chemical Research, Slovak Academy of Sciences. We are especially grateful to Prof. T. J. Kemp, the editor of this English translation, for making innumerable constructive and helpful comments concerning both the language and the topic.

Authors

I. INTRODUCTION

Systematic investigation of the chemical reactions of polymers was initiated by the desire to improve the properties of natural materials, by such processes as the tanning of leathers the vulcanization of rubber and the etherification and esterification of cellulose. It seemed at first that the discovery of synthetic polymers would relegate the techniques of polymer modification to the periphery of research and applications. This did not happen however, since it was soon found that novel synthetic polymers may be produced by transformation reactions of macromolecules which feature properties hard to obtain by simple polymerization reactions of monomers.

Currently the topic of the chemical reactions of polymers as developed within macromolecular chemistry is relatively well-defined. It encompasses all reactions where macromolecular compounds act as starting materials. Specific transformations of functional groups in polymers taking place with minimal change in the degree of polymerization but with a distinct change of physicochemical properties are termed "polymer-analogous reactions". In reactions leading to the destruction of polymers, the usually unwanted scission of the macromolecular backbone occurs by some physical perturbation or by attack of chemical reagents. The syntheses of graft and block copolymers and the reactions of polymer molecules leading to crosslinked polymers are also included in this overall category. The latter types of reaction require a relatively minor chemical intervention in the parent macromolecules and thus represent a modification of polymer structure at a quasi-supermolecular level. Minor and even reversible structural changes also take place in polymer catalysts. The reduced extent of chemical modification of the chains in these various processes does not imply however that such reactions are less important.

The types of chemical reaction of polymers outlined are universal for both synthetic and biological macromolecules. For example, the physiological fate of proteins in vivo is determined by a sequence of modification reactions. These are largely the enzymatically-controlled chemical modifications of the polypeptide backbone and of the side groups, such as the formation of the disulphide linkage between the segments of a protein chain or its disruption, the linkage of carbohydrates to their side groups, the formation of crosslinking bridges between macromolecules or the scission of chemical bonds in the main chain. During the biosynthesis of collagen as many as ten stages of modification of the initial macromolecules have been identified.

Several polymer-analogous reactions such as preparations of cellulose derivatives, the chlorosulphonation of polyethylene, the saponification of poly(vinyl acetate) to poly(vinyl alcohol), the chlorination of poly(vinyl chloride), and the functionalization of polymer gels achieved industrial importance long ago. In these reactions, partial or even almost complete exchange of the original functional groups for a new ones takes place, and the final product has the character of a copolymer. Sometimes only a small number of the functional groups is incorporated into the macromolecule, displaying then specific physical, chemical or biochemical effects (to form polymer stabilizers, catalysts and drugs). The linkage of amino, sulphonato-, carboxylato- and phosphonato-groups to the polymer backbone is widely used practically in the production of ion-exchange resins. The alternative route involving the complete transformation of the functional groups, namely, the polymerization of the corresponding monomers, is sometimes complicated or even impossible.

The technology of processing polymer materials may also interfere with the direct synthesis of a polymer with desirable properties. For example, rubber products are initially formed to the required shape which is then fixed by subsequent vulcanization. Crosslinking is unavoidable in this case since natural rubber is a linear polymer. The crosslinking of synthetic elastomers also takes place as a subsequent process even though in this case the network structure could, in principle, be formed during their synthesis. However, the crosslinks fix the macromolecular structure to such a high degree that the final product cannot be fabricated by extrusion or moulding but only by cutting, latheing or other mechanical treatment. Economic grounds mean that such processing can be utilized only rarely.

The chemical transformations of polymers are aimed not only at broadening the spectrum of properties of conventional polymers but also to prepare the polymers for some specific purpose such as energy transducers, selectively permeable membranes, media of information storage, etc. This latter aspect, the preparation of speciality polymers and their applications, is very promising for the near future.

Knowledge of the mechanism and, in general, of the pattern of chemical reactions of functional groups in polymers enable us to control both desirable and unwanted processes in polymer materials. Finally, one has to proceed step-by-step in the clarification of the differences in the chemical transformations of biomacromolecules in the normal and pathological states of living bodies. Our understanding of these vital processes would have far-reaching and beneficial consequences for mankind.

References

1. Chemical Reactions of Polymers, FETTES, E. M. (Editor), Interscience, New York 1964.
2. CASSIDY, H. G., KUN, K. A.: Oxidation-Reduction Polymers, Interscience, New York 1965.
3. Chemical Transformations of Polymers, RADO, R. (Editor), Butterworths, London 1971.
4. Reactions on Polymers, MOORE, J. A. (Editor), REIDEL, D. Publ. Co. Dordrecht 1973.

II. CHARACTERISTIC FEATURES OF CHEMICAL REACTIONS OF MACROMOLECULES

It is a common assumption in macromolecular chemistry that the reactivity of a functional group does not depend on the size of the molecule to which is attached. This assumption is essential for the kinetic analysis of polymerization reactions since a constant rate is postulated for the reaction of monomer with a growing chain regardless of its gradually increasing mass. As to the resulting macromolecules, it has indeed been verified experimentally that a functional group attached to a chain reacts with approximately the same rate as the same group present in small molecules. The functional group can be a part of each repeat unit of the polymer chain or may form only an occasional reaction site in the macromolecule.

At first glance it may thus appear that both the reactivity of a functional group in a polymer chain and the composition of the reaction products will be the same as in small molecules and do not require special study. It is known however that in reality numerous exceptions from the above rule exist, denoted by the general term the "polymer effect". The level of reactivity of the functional groups in macromolecules is connected with a more condensed and differently organized systems of reactants. The reactions of a macromolecular compound and of its low-molecular-mass analogue will probably follow a similar course when the reaction system is homogeneous, with a sufficiently large diffusional mobility of all reactants, intermediates, and products and when only one type of functional group reacts. However, the low molecular compound selected for comparison should mimic as far as possible the steric conditions on the polymer chain. The reduced diffusional mobility of macromolecules and steric hindrance to the approach of reagents to functional groups are frequent causes of a lower reaction rate in a macromolecular system. The polymer effect sometimes manifested in an enhanced reaction rate of a polymer reagent is a consequence of the specific primary, and especially secondary and tertiary structure of the polymer chain.

The largest reduction in reactivity of a functional group is observed in the bimolecular reactions of mutually noncomplementary macromolecules. The reduced rate follows from steric and/or electrostatic hindrance to the reaction centre by the macromolecular backbones P_1 and P_2. We can express this formally by the scheme

$$\begin{pmatrix} P_1 & & P_2 \\ -A & + & B- \end{pmatrix} \rightarrow \begin{pmatrix} P_1 & & P_2 \\ -C & + & D- \end{pmatrix}$$

which should indicate both obstruction to the approach and to suitable juxtaposition of the reacting groups A and B. It is evident that arrangements with more accessible functional groups created, for example, by conformational transitions, should be more reactive.

In the reaction between a macromolecular reagent and a low molecular compound, again, the shielding of a functional group by the chain may retard reaction. However, an opposing influence is also possible, namely the preferential inclusion of compound B into functional group A. This formal scheme can be applied to enzymatic reactions. In this case, however, the catalytic effect of the rest of the macromolecule involves not only the preferential inclusion of the reactant but also its proper orientation and cooperative activation by neighbouring groups.

The problem of the influence of the macromolecular chain is completely different in a monomolecular reaction

$$\begin{pmatrix} P & & P \\ -A & \rightarrow & -C \end{pmatrix}$$

when the reaction can be faster or slower than a reference low-molecular reaction, depending on the compensation of the activation factors from the environment and the reduced mobility in the transition state.

Macromolecular reactions can be classified as monofunctional and multifunctional depending on the number of functional groups on the chain entering in one reaction step. Monofunctional reactions includes many polymer-analogous reactions such as methylation, acetylation, halogenation or hydrolysis of a side group in the chain. In multifunctional reactions, generally two neighbouring groups are mutually transformed as in the formation of polymer anhydrides by cyclization, for example from poly(acrylic) acid

A lower degree of conversion is achieved in multifunctional reactions as compared to the corresponding reactions of low-molecular-mass compounds. The difference results from the random siting of functional groups during reaction;

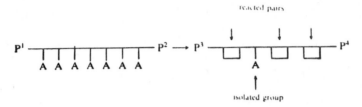

isolated groups lacking an active neighbour cannot react. Multifunctionality is also typical of enzyme catalysis. By means of the multifunctional cooperation of the functional groups and the reacting substrate S

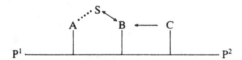

the substrate is bound by intermolecular forces to group A so that the time of contact increases and a suitable orientation of the neighbouring reacting group B, activated by group C is ensured. The rigidity of the carrier of the functional groups promotes the selectivity of reaction.

The specificity of a multifunctional reaction can be enhanced by the mutual complementarity of segments of two rigid macromolecules due to the physical interaction of groups A and B

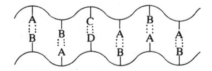

Due to this attractive interaction, a close approach of the reacting groups C and D is achieved and spontaneous reaction takes place.

A comparison of the reactivity of a given functional group in polymers prepared by various methods is affected by the distribution of the defect centres along the macromolecular chain that is hard to detect by conventional analytical techniques but which could be significant for the reactivity of the chain. The various low-molecular additives, impurities, and residues of initiation systems could all have similar effects on reactivity as defects in the constitution of

16

individual mers. These effects can be minimised, however, by careful preparation of samples.

The reactions of macromolecules are also influenced by interchain packing and supermolecular organization, such as the occurrence of several solid-state phases, which have no analogue in low-molecular-mass compounds. One consequence of variable morphology of a given polymer in the crystalline state is that the rate of a reference reaction may change and thus contribute to the disparity of kinetic results from different laboratories.

The majority of these effects of polymer environment on the reactions act cooperatively and thus the elucidation of the reactivity of a functional group is not always clear-cut and definitive.

A. EFFECT OF NEIGHBOURING GROUPS

The influence of neighbouring groups on the reactivity of a functional group is well-known in organic chemistry and it is pronounced in polymer chains. The nearest neighbour (an electrophilic or nucleophilic group) is usually the most important and effects an increase or reduction in the reactivity of an isolated functional group. A well known example of the neighbouring group effect in polymer systems is the technically important base-catalysed hydrolysis of poly-(vinyl acetate) to poly(vinyl alcohol). In this case, the reactivity of the ester group is enhanced by neighbouring hydroxyl groups formed during the reaction. Two hydroxyl groups cooperate in the transformation of poly(vinyl alcohol) to poly(vinyl acetal) by the addition of aldehyde (usually formaldehyde or butyraldehyde) in acidic solution

The reaction proceeds almost completely with a surprisingly small residue of unreacted groups. This is in apparent contradiction to statistical calculations, which predict that about 13 % of groups should not react when identical reactivity is assumed. Experimental measurements of irreversible reactions such as the dechlorination of poly(vinyl chloride) actually support this theoretical prediction. However, the formation of poly(vinyl acetal) is a reversible reaction and almost all functional groups can react by successive mutual exchange [1].

The chain elements in the vicinity of a functional group can affect its reactivity in some other ways. *Steric shielding* diminishes reactivity by blocking the access of a reagent to the active centre and by affecting the geometry and the

17

probability of formation of the transition state. This effect is pronounced especially when the neighbouring group contains a quaternary carbon atom such as in methacrylate polymers and poly(α-methylstyrene). Steric effects are also demonstrated by the different degrees of sulphonation of poly(vinyl alcohol) and its derivatives: here the dependence on the substituent R

$$-CH_2CHCH_2CHCH_2CH-$$
$$\quad\quad OR\quad\ OR\quad\ OR$$

is maximised when R is hydrogen or methyl, is lower for phenyl and still lower for the naphthyl group.

Electrostatic or polar influences may hinder or facilitate the close approach of charged substituents or impede the introduction of an electronegative substituent in the neighbourhood of one already present, as in the case of chlorination. Electrostatic interactions retard the alkaline hydrolysis of polymethacrylamide and even stop it completely at about 70% conversion. The hydroxyl anion is electrostatically repelled from the methylacrylamide units by two methacrylate ions:

$$\begin{array}{ccccc} & CH_3 & & CH_3 & & CH_3 \\ -C & -CH_2 & -C & -CH_2 & -C- \\ COO^- & & CONH_2 & & COO^- \end{array}$$

In the hydrolysis of copolymers of acrylic acid and 4-nitrophenyl methacrylate, the dissociated carboxyl groups act on the ester group, and 4-nitrophenol is eliminated faster than in the corresponding reactions of low-molecular-mass compounds [2].

The dependence of macromolecular reactivity on stereoisomerism in the vicinity of the functional group represents a special example of the neighbouring group effect. In vinyl polymers this means that the chemical reactivity of groups depends on the type of microtacticity. For example, the syndiotactic isomer of

18

poly(vinyl alcohol) is more reactive in the transformations into poly(vinyl acetate) and poly(vinyl acetal). In another example of a pronounced neighbouring group effect, the degradation of poly(vinyl chloride), the dependence of the kinetic parameters of degradation was related to the tacticity of the initial sample.

In the course of reaction, the conformational effect may increase in importance since, with the progressive modification of the chain, certain conformations may be favoured which differ in reactivity from the initial conformations.

The nearest-neighbour effect on the reactivity of functional groups can be expressed quantitatively by a statistical treatment of the reaction kinetics of polymer-analogous reactions. The degree of conversion of a reaction in the chain depends on the three rate constants k_0, k_1, k_2, when the assumption is made that groups A transform to groups B irreversibily by a first order reaction. The rate constant k_0 characterizes the transformation of A to B in a sequence (triad) AAA, the constant k_1 in triads AAB or BAA (with one transformed group B in the neighbourhood of group A) and the constant k_2 in the triad BAB with two transformed neighbouring groups. The formal kinetic theory of the process [3] provides the distribution of reacted and unreacted units in the chain, or in other words, the compositional heterogeneity of the resulting copolymer.

The magnitude of the neighbouring group effect is determined by the ratio of the above three rate constants. There is no effect when all three constants are identical, and the groups A and B are randomly distributed in the resulting product. When the constant k_0 is substantially smaller than the other two, $k_0 \ll k_1 \leq k_2$, the neighbouring group accelerates the reaction and blocks of group B are formed in the copolymer chain. When on the other hand, $k_0 \geq k_1 \geq k_2$, the transformed group retards the reaction and isolated groups B will predominante in the macromolecule.

The formal kinetic theory of the polymer-analogous reactions is well-established but unfortunately much less attention has been paid to the reverse problem, i. e. to the actual determination of the rate constants k_0, k_1, k_2 from experimental data. Comprehensive data exist for the individual rate constants for the reactions of polymethacrylate esters (*Table 2.1*). A pronounced enhancement of the reaction rate by the neighbouring group effect is apparent in most cases and a marked influence of tacticity is evident in the case of poly(diphenylmethyl methacrylate); note that the syndiotactic configuration of this polymer features identical reactivity for all its ester groups in a weakly alkaline medium.

The kinetic theory of polymer-analogous reactions combined with a knowledge of the individual rate constants enables calculation of the distribution of monomer units in the chain. The heterogeneity in the composition of copolymers is important in the evaluation of the resulting chemical, physical and mechanical properties of the products of polymer-analogous reactions. For

Table 2.1   The effect of neigbouring groups on the rate constants of polymer-analogous reactions of polymethacrylic esters at 145 °C [3, 4]

Reaction	Tacticity	Medium	$k_0 \; 10^{-4}$ min^{-1}	$k_0 : k_1 : k_2$
Hydrolysis of poly(methyl methacrylate)	iso	KOH (0.2 M)	90	1 : 0.4 : 0.4
	syndio	KOH (0.2 M)	5.8	1 : 0.2 : 0.05
	syndio	KOH (0.2 M)	1.9	1 : 0.7 : 0.7
	iso	pyridine-water	5.3	1 : 8 : 100
	syndio	pyridine-water	1.1	1 : 2.5 : 3.4
Hydrolysis of poly(diphenylmethyl methacrylate)	iso	pyridine-water	0.3	1 : 20 : 100
	syndio	pyridine-water	0.3	1 : 1 : 1
Hydrolysis of poly(phenyl methacrylate)	iso	pyridine-water	6	1 : 40 : 1 000
	iso	dioxan[a]	0.15	1 : 55 : 100
	iso	dioxan[b]	2	1 : 18 : 65
	hetero	dioxan[a]	5	1 : 2 : 10
Methylsulphonylmethyl-lithium with poly(methyl methacrylate)	syndio	DMSO-benzene[c]	16.5[d]	1 : 0.055 : 0
	iso	DMSO-benzene[c]	78.3[d]	1 : 1 : 0.003

a) 80 °C. b) 100 °C. c) 25 °C.
DMSO: dimethyl sulphoxide. d) $k \; 10^{-4} \; (\text{dm}^3\text{mol}^{-1}\text{s}^{-1})$

example the sequential distribution and the compositional heterogeneity have been determined in a polyethylene chain after chlorination and compared with the results of fractionation of the copolymer. The best agreement between the calculated and experimental kinetic curves for the chlorination of polyethylene in chlorobenzene at 50°C was found [3] for the ratio of the rate constants of $k_0 : k_1 : k_2 = 1 : 0.35 : 0.80$. A similar ratio was found also for the reactions of higher linear alkanes and cycloalkanes. It seems therefore, that the chlorination of polyethylene is an example of a reaction where conformational and other effects relevant to longer chains do not influence the reactivity of the structural unit.

The neighbouring group effect is also important in some degradation reactions. In the thermal degradation of poly(vinyl chloride), hydrogen chloride is eliminated but the integrity of the polymer backbone is retained. The resulting double bonds activate the further elimination of HCl from a neighbouring group

$$—CH_2CH{=}CHCHCH_2CH— \;\longrightarrow\; —CH_2CH{=}CHCH{=}CHCH—$$
$$\quad\quad\quad\quad\quad | \quad\quad | \quad\quad\quad\quad\quad\quad\quad\quad\quad\quad |$$
$$\quad\quad\quad\quad\quad Cl \quad\; Cl \quad\quad\quad\quad\quad\quad\quad\quad\quad Cl$$

The elimination of the substituent also takes place on thermal degradation of complex polyacrylates and polymethacrylates with the release of alkene, as in thermolysis of poly(*tert*-butyl methacrylate)

$$
\begin{array}{c}
\text{CH}_3 \\
(-\text{CH}_2\overset{|}{\text{C}}\text{H}-)_n \\
\overset{|}{\text{C}}=\text{O} \\
\overset{|}{\text{O}}\text{C(CH}_3)_3
\end{array}
\quad \longrightarrow \quad
\begin{array}{c}
\text{CH}_3 \\
(-\text{CH}_2\overset{|}{\text{C}}\text{H}-)_n \\
\overset{|}{\text{C}}=\text{O} \\
\overset{|}{\text{O}}\text{H}
\end{array}
+ n\,(\text{CH}_3)_2\text{C}=\text{CH}_2
$$

in which the neighbouring carboxyl group activates the elimination of isobutylene.

A special case of reactions influenced by nearby functional groups is given by those where the polymer acts as a catalyst including, especially, enzymatic reactions. In the latter case however, the complete organization of the chain is responsible for the activation of interacting groups and we will consider this important problem separately.

B. INFLUENCE OF THE MEDIUM ON SOLUTION REACTIONS

Much information on chemical reactions comes from investigation in solution. Just as with the reactions of small molecules, the reactions of polymers also are influenced by the nature of the solvent, with some of their specific features being related to the shape adopted by the macromolecules in solution. Most synthetic polymers and also some biological polymers form a random coil in solution, which represents the set of all conformations with end-to-end distances fitting a Gaussian distribution. Under normal conditions biological (and some synthetic) macromolecules build up in solution fully- or partially- ordered structures such as helices. These can be converted into a random coil structure by a change of solvent, pH, temperature, etc. The size of a statistical coil and its "swelling" depends on macromolecule-solvent interaction and thus on the so-called "thermodynamic quality" of the solvent. The coils of a given polymer are expanded more in some solvents and less in others. An increase in concentration of the polymer in solution brings about a compression of the coil (i.e. a reduction of its effective hydrodynamic volume) and of its mutual interpenetration with other coils.

Solvent effects may considerably complicate the treatment of the chemical reactions of polymers. The reactions seldom proceed to completion due to the reduced accessibility of the functional groups or to the limited solubilities of individual components of the reaction mixture. A solvent, in which the resulting transformed polymer precipitates is usually chosen on practical grounds. The kinetic study of reactions featuring polymer-polymer interaction is especially complicated. Even if the system satisfies the condition of mutual compatibility in solution and no phase separation occurs, the reaction products increase the solution viscosity and retard the reaction.

The medium effect in polymer reactions may also differ from that for small molecules as regards the microscopic situation near the reaction centre. Polymer backbones contribute to the effective medium (solvent) of the functional group and the properties of the "local" solvent may differ substantially from those of neat solvent. When the reaction is sensitive to the solvent polarity, the effect of the concentration of the local solvent or microenvironment can be responsible for an "anomaly" in the reactivity of a polymer reagent.

Interactions of a polymer with a reagent of low-molecular-mass may also change the concentration of small molecules in the polymer domain, which naturally affects the reaction rate. The use of solvents, particularly of multicomponent ones, with a different affinity towards polymer molecules than to a small-molecule reagent makes the situation even more complex. Due to the preferential solvation of the polymer, the solvation sphere of the chain is enriched preferentially by one component of the mixture.

This and similar effects are particularly prominent in polyionic reactions [5]. When polyions react with charged small molecules of the same sign, the reaction is impeded by electrostatic interactions due to the high charge density on the polyion and conversely, reactions between oppositely charged ionic compounds are accelerated. The addition of electrolytes makes both these trends less pronounced as a consequence of the primary salt effect. The reactions of 2-bromoacetate ion provides an example, either with a polyanion (partially dissociated poly(methacrylic acid))

or with a polycation (partially protonated poly(4-vinylpyridine))

22

In the first case, the reaction proceeds slowly, although the analogous reaction with uncharged bromoacetamide is very fast. In the case of the polycation, electrostatic attraction of the reagents increases the reaction rate, although this increase can be inhibited by the addition of neutral salt.

A reduction of reaction rate with a decrease of charge density is frequently observed in reactions of polyanions and polycations with neutral organic reagents which is apparently associated with the preferential solvation in these systems. A relatively nonpolar small-molecule reagent is progressively excluded from the vicinity of the chain on increasing the solvent polarity. It is possible to eliminate or suppress this decrease by employing a less polar solvent, with a polarity closer to that of the low-molecular-mass reagent.

The influence of the medium either on the effective dimensions of the coil or on the polarity of the microenvironment of the reaction centre complicates interpretation of the dependence of the kinetic parameters of the reaction on the structure of the chain. For example, when comparing the kinetic parameters of vinyl polymers of different tacticity one should bear in mind that a change in configuration may also induce change in polymer-solvent interaction.

C. INTRACHAIN REACTIONS

Intrachain cyclization is a highly characteristic example of the reactions of functional groups sited on the same chain. Cyclic molecules with small (mainly six-membered) and medium-sized rings are well-known from the organic chemistry of small molecules. Their analogues also exist in polymers, formed by cyclization between two functional groups in close proximity. Chain flexibility is a necessary prerequisite for reaction between distant groups and for formation of large rings containing tens or even hundreds of monomer units. Three types of

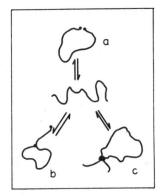

Fig. 2.1. Schematic representation of the cyclization of random coil macromolecules
(*a*) end-to-end cyclization, (*b*) end-backbone cyclization, (*c*) intrachain cyclization

cyclization process can be visualized in a macromolecule (*Fig. 2.1*); (i) end-chain cyclization, (ii) end-backbone and (iii) intra-backbone cyclization. End-chain cyclization in particular is a topic of considerable interest.

Cyclization reactions in dilute solution can be classified as *activation-* or *diffusion-controlled reactions* [6]. The former are reactions of functional groups with considerable activation energy and therefore with a low value of the rate constant. The rate of reaction is determined in this case by the probability of finding a chain in a conformation favourable for reaction, where both functional groups (located for example at the chain end) are sufficiently close. Intramolecular reactions of this type provide information on equilibrium conformational properties of chains and are termed as conformationally-controlled reactions.

In diffusion-controlled cyclizations, the reaction of the functional groups is very fast, and consequently, the rate-determining step becomes the frequency of collisions between active groups. This type of process provides data on the rate of those conformational changes which potentially lead to the contact of groups, i.e. information on the dynamic behaviour of the chain. The mutual approach of the two reactants is the faster the larger is the sum of their diameters and their diffusion coefficients. The decisive criterion characteristic of diffusion--controlled reactions is the sensitivity of their rates to the viscosity (η) of the medium. The Debye equation applies to the rate constants of these reacti ns

$$k_{\text{diff}} = 8\, RT/3000\eta$$

where k_{diff} is expressed in $\text{dm}^3\,\text{mol}^{-1}\,\text{s}^{-1}$ and the viscosity in mPa s (the equivalent of cP units).

Turning briefly to reactions controlled by the activation energy, i.e. which are conformationally controlled. Cyclization to give small rings occurs mainly at the expense of the deformation of bond angles, and unfavourable steric interactions increase the potential energy of the ring (to give ring strain). The length of the polymer chains enables the system to realize a large number of chain conformations in which factors unfavourable to the formation can be avoided. The probability of cyclization is determined by the conformational statistics, which in turn depend on the energy and entropy differences between the stable isomers for each single bond in the chain. However, the possibility of intrachain cyclization is not limited to flexible polymers only, and may occur with a rather conformationally rigid polymer such as deoxyribonucleic acid (DNA). Cyclic DNA is well-characterized and seems to play an important role in gene replication. Even the catenane (metal-chain) type of linking of DNA rings has been observed.

The probability of encounters of functional groups in random coils where their end-to-end distances (r) follow a Gaussian distribution is given as

$$W = 4\pi r^2[3/(2\pi\langle r_0^2\rangle)]^{3/2}$$

The mean square of the end-to-end distances $\langle r_0^2\rangle$ is directly proportional to n, the number of monomeric units in the chain between the reacting groups; hence, the expression predicts a decrease of probability of cyclization with $n^{3/2}$.

Numerous data are available on the kinetics of cyclization reactions, especially for macrorings with a variable number of methylene units. Compounds of the type $A\text{---}(CH_2)_n\text{---}B$ form the rings $(CH_2)_n\text{---}D$ where A and B and D could be, for example, Br, NH_2 and NH groups, respectively. An intrachain reaction of similar type is the hydrolysis of substituted poly(N-methyl-glycine) with a p-nitrophenyl ester group on one end and a pyridyl group on the other [7]

$$\text{N}\bigcirc\text{---CH}_2\text{NH}\text{--(COCH}_2\text{N(CH}_3\text{))}\text{---}_n\text{COCH}_2\text{CH}_2\text{CO O}\bigcirc\text{---NO}_2$$

The hydrolysis is decelerated by increasing the length of the chain (*Fig. 2.2*) in accordance with the above relationship for the probability of a chain-end encounter.

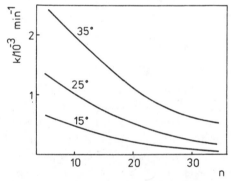

Fig. 2.2. Variation of the rate constant for the intramolecular hydrolysis of terminally substituted poly(N-methylglycine) with the number of mers n for three temperatures [7]

Of special interest are situations where the end-group can react with monomer units in the chain backbone, as in the following example

$$x+y=m-1$$

$$\longrightarrow \text{products}$$

where the electronically excited carbonyl group in benzophenone represents the reactive end-group. The intramolecular abstraction of hydrogen should occur at each methylene group which can be approached by an oxygen atom of the excited group. It turned out [6] that individual reaction sites along the chain differ in reactivity, thus the first nine CH_2 groups are almost excluded from involvement in reaction.

Finally, we can mention an example of diffusion-controlled intrachain reactions [8] with naphthalene (Nf) and its anion-radical linked to each end of a methylene chain of variable length in the compound $Nf-(CH_2)_n-\dot{N}f^-$. The exchange of an electron localized on a naphthalene is known to be extremely fast and consequently the rate-determining step is the diffusion-controlled chain--end encounter, since each collision should bring about electron transfer. Photophysical measurements using a suitable chromophore at the chain end are mostly used for the determination of the rates of end-group collisions. The rate constant in diffusion-controlled cyclization is a measure of the dynamic flexibility of the chain as determined by the barrier to internal rotation about the single bonds in the polymer backbone. It is not surprising therefore, that polyoxyethylene or polydimethylsiloxane with end groups cyclize faster than the analogous derivatives of polystyrene with the same chain length [6].

D. INTERCHAIN REACTIONS

Interchain reactions represent the third group of bimolecular reactions of polymers in addition to reactions of polymers with small molecules and to intrachain cyclization. The kinetics of reactions of two polymers of types P—A and P—B solution exhibit several features common with those of intrachain cyclization. In this case activation- and diffusion- controlled reactions can be differentiated. Functional groups may be localized either randomly along the chain or at some specific site such as at the chain end. As indicated previously, this type of reaction is exceptionally strongly affected by solvent-macromolecule interactions.

The reactivity of polymers in an activation-controlled reaction is usually evaluated by comparison with an analogous small-molecule reaction. We can take as an example the reaction of benzoyl chloride with 4-benzoyloxy-2,6-dinitrophenol $(R = C_6H_5)$

26

which proceeds in the presence of pyridine catalyst with a rate given by the rate constant k_0. The same functional groups, chloride and nitrophenol, can be bound separately on polystyrene chains with variable molecular mass M either by random substitution or at the chain end [9]. It transpires that the rate constant of the polymer reaction k approximates to k_0 in sufficiently concentrated solutions when the polymer coils are substantially interpenetrating. The ratio k/k_0 is reduced to $0.1 — 0.2$ when reaction proceeds in dilute solution. It was found moreover, that this reduction is more pronounced for substituents localized randomly along the macromolecule than when they are at the chain ends. Again, it is evident from this example that functional groups situated along the chain may normally be considered as equivalent, but in some reactions certain locations of the functional groups may display enhanced reactivity.

The question of the dependence of the rate of interchain reactions upon chain length is still unresolved. We have noted earlier that for the majority of reactions of polymers with low-molecular-mass compounds the rate does not depend on the chain length. For example, no change of reaction rate was observed for the reaction of dansyl chloride (5-dimethylamino-1-naphthalene sulphonyl chloride) with poly(ethylene oxide) terminated with a primary amino group when the molecular mass of the polymer was changed from 10^3 to 10^5. The decrease of the ratio k/k_0 to less than unity in interchain reactions is caused by the "exclusion" of one macromolecular reagent from the space occupied by the coil of the second reagent. Since the coil size increases with molecular mass M, the ratio k/k_0 should decrease with larger values of M and with thermodynamically better solvents. In some cases as for example in the bimolecular reactions of functionalized polystyrene macromolecules mentioned in the previous paragraph, a decrease of k/k_0 with M was actually observed [9]. Some other authors, however, contest the existence of a "kinetically excluded volume" since k/k_0 was found to be independent of M in the interchain reaction of polyethylene macromolecules terminated with chlorosulphonyl and primary amino groups [10].

Activation-controlled interchain reactions in solution are closely related to the formation of equilibrium *interpolymer complexes*. Just as with polymer reactions the degree of association and the equilibrium constant K of the polymer-polymer complex should decrease with increasing M due to the excluded volume. Some of the most important complexes are those formed in the cases when one component is a polyacid and the other a polybase, such as complexes of poly(acrylic acid) or poly(methacrylic acid) with poly(ethylene oxide). The outstanding feature of these complexes, i.e. the cooperative character of macromolecule-macromolecule interactions, is evidenced by an abrupt change of properties over a narrow interval in an applied perturbation or set of conditions.

The reactivity of functional groups in polymer-polymer complexes depends on the conditions of complex formation. A poly-complex formed by mixing of diluted macromolecules in a common solvent exhibits more defects (*Fig. 2.3*) than a complex formed by the polymerization of one component on a matrix of the second component. The structural inhomogeneity of a polycomplex results in the differential reactivity of functional groups located in the various regions.

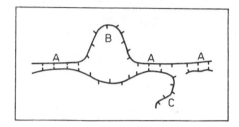

Fig. 2.3. Representation of the structural heterogeneity of a polycomplex of two interacting macromolecules with the coexistence of ordered regions (*A*), loop defects (*B*) and loose ends (*C*)

For example, the intrachain cyclization of poly(acrylic) or poly(methacrylic) acid with the formation of anhydride units proceeds solely in the defect regions of complexes of above polyacids with poly(ethylene oxide), poly(vinyl-pyrrolidone) or polyacrylamide. In contrast, intermolecular amidation is possible only in regions of intimate contact of functional groups of both types of polymers. However, the latter reaction comes rapidly to a stop due to the rigidity of the polycomplexes induced by polyamidation [11].

Biological macromolecules provide numerous cases of associations of pairs of macromolecules, mainly by means of hydrogen bonds, of which the double helix of deoxyribonucleic acid is a prominent example. The same principle can also be used in the field of synthetic polymers for enhancement of the compatibility of two types of polymer. Introduction of 4-vinylpyridine units into polystyrene and methacrylic acid units into poly(methyl methacrylate) induces hydrogen bond formation between these groups and enhances the mutual interpenetration of both types of coil in solution. A similar effect can be accomplished by polyanion-polycation interaction as, for example, between sulphonated polystyrene and polyacrylate, to which a small amount of quaternary pyridinium units has been linked.

The mechanism of diffusion-controlled interchain reactions is more complex. The diffusional process can be divided in this case into two independent steps. The first, translational diffusion of the centre of gravity of the polymer chain, is called the macro-Brownian process and the second translational and rotational diffusion of segments with functional groups is denoted as the micro-Brownian movement (*Fig. 2.4*). The classical example of a diffusion-controlled reac-

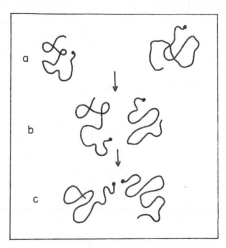

Fig. 2.4 Various stages in the diffusion-controlled reaction of two macromolecules. The chains are brought together by translational diffusion (a → b), they are suitably oriented by segmental diffusion (b → c) after which the functional groups can react.

tion of polymers is bimolecular termination in radical polymerization, where diffusion of the end-segment of the chain is the rate-determining step. Apart from viscosity and temperature, the rate constant k_{diff} for the case where segmental (micro-Brownian) diffusion is the rate-determining step, depends on the polymer-solvent interaction (i.e. on the degree of expansion of the coil α) and on the molecular mass of polymer; this may be expressed formally as

$$k_{diff} = A(\alpha)B(M)C\left(\frac{1}{\eta},T\right)$$

Theoretical expressions have been derived [13] for the individual terms A, B, C. The rate constant for a functional group on the chain end is predicted to be

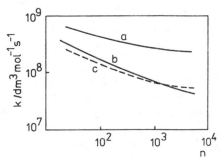

Fig. 2.5. Graphical plot of the dependence of the rate constants for the recombination of poly(ethylene oxide) radicals versus the number of mers n

Theoretical curves for (*a*) end-radicals, (*b*) randomly localized radicals and (*c*) experimental data [13]

29

slightly higher than for a group located randomly in the chain. *Fig. 2.5* shows theoretical and experimental curves of the rate constant k_{diff} for the recombination of poly(ethylene oxide) radicals generated by pulse radiolysis in aqueous solution. It is seen that k_{diff} decreases with increasing molecular mass of the polymer. The experimental data are best fitted to the curve on the assumption of random distribution of radical centres along the chain. Apparently, transfer reactions of radicals inside the coil contribute to the physical diffusion.

The above equation shows that the variation in the type of solvent does not influence k_{diff} through a change of viscosity only, but also by alteration of its thermodynamic properties. At the same time, the more rigid is a chain the smaller is k_{diff}. Hence, a reciprocal relation between k_{diff} and η is observed in diffusion-controlled interchain reactions only when the solvents under comparison have similar thermodynamic qualities.

The diffusion-controlled reactions of two macromolecular reagents in solution are closely related to the kinetics of chain scission in solution insofar as these are mutually reverse processes. If the kinetics of random scission are not activation-controlled then micro- and macro- Brownian kinetics should also be considered in this case. The process of chain scission is reversible as long as the respective products are still localized in the solvent "cage", and is terminated only after diffusion of the chain fragments from the cage. The conformational flexibility of the chain and the local viscosity of the medium determine the relative importance of translational and segmental diffusion in the overall process. These factors also contribute to the dependence on chain length of the rate constant of random scission which has been observed, for example, for atactic polystyrene under the action of NO_2 in dioxan or in the solid state [14].

E. REACTIVITY OF SUBSTITUENT GROUPS IN THE GEL PHASE

The reactivity of functional groups attached to crosslinked polymers is specific in several ways, differing often from their reactivity in solution. The partially retained mobility of functional groups and the presence of solvent in the gel reduce those restraints to reactivity typical of the solid state. The broad application of functionalized crosslinked polymer supports began following the successful application of this method in the multistep synthesis of polypeptides with a controlled sequence of structural units.

Reactivity in the gel phase is closely associated with the structure of the carrier (15, 16). Depending on the type of crosslinking, the density of crosslinks and the solvent, the carriers can be roughly classified within the two extreme cases shown in *Fig. 2.6*. The carriers of the *"gel"* type are formed from the

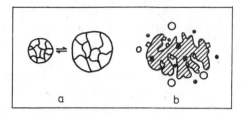

Fig. 2.6. Crosslinked polymer carriers of gel type

(*a*) gel character achieved by swelling, (*b*) macroreticular gel type with a solid porous structure with surface (□) and internal (■) functional groups of different accessibility towards small (●) and large (○) molecules

less-densely crosslinked homogeneous gel prepared in a good solvent and exhibit a variation in their degree of swelling with the diluent used, a small pore diameter and a large effective surface. In contrast, the heterogeneously crosslinked *"macroreticular"* carriers involve regions with an aggregated, partially-ordered network structure which is not perturbed by the action of diluents. The characteristic feature of macroreticular carriers is the presence of large pores, however their total effective surface area may be small.

Reactions of functional groups linked to the gel proceed usually more slowly than the analogous reactions in homogeneous solution. The porosity, density of crosslinking of the carrier and reagent size impose restraints on diffusion in the gel to such a degree that the kinetics of reaction become diffusion-controlled. Generally, the path of a reagent from the surface of a particle to an interior functional group should be the shorter the smaller the size of the gel particle. Similarly, the diffusional track becomes longer and more tortuous with an increase of the network density in a particle. Hence, the reaction rate should increase both with a reduction of particle size and with the degree of crosslinking. However, the micro-and macroporosity complicate this simple reasoning. In simple terms, macropores are pores already existing in dry gel particles while some micropores may be formed additionally after contact with a solvent and swelling of the particle. The transport of reagents in the macropore carrier proceeds through the swelled polymer matrix as well as through macropores filled by solvent where the diffusional rate is similar to that in pure solvent. In the polymer matrix, microscopic fluctuations exist in the homogeneity of the crosslinking, with regions of less dense network existing where the restraints to diffusion are similar to those found in concentrated solution.

Thus, the heterogeneous structure of the carrier may modify the local diffusional coefficients of reactants and products by several orders of magnitude. This phenomenon contributes to the kinetic nonequivalence of individual groups linked to the gel. The dependence of the kinetic parameters of reaction on the particular locality of functional groups invalidates the standard assumption of solution kinetics. The dispersion in reactivity of functional groups can be exemplified by the linking of isoleucine to a gel-attached polypeptide chain where it was found [17] that 90% of all groups reacted within 10 minutes but a further 5 % only after two hours.

An important role in the kinetics of a reaction in the gel phase is played by *solvation effects*. The polarities of the macromolecular network and the diluent determine the degree of swelling, and they should be matched optimally. Therefore, nonpolar polystyrene gels have been replaced by more polar ones such as polydimethylacrylamide gels for reactions of biopolymers in aqueous media. The accessibility of functional groups depends on the type of spacer by which the active group is bound to the crosslinked matrix. Longer and more flexible spacers make the functional group more accessible to reagents [18]. Finally, we should point out that the above characteristic features of reactions in the gel phase apply equally well to functional groups taking part in reactions as catalysts.

F. STRUCTURAL ORDER AND REACTIVITY IN THE SOLID STATE

The reactivity of polymers in the solid state is profoundly affected by regularity in the packing of the chains in the solid state as determined, for example, by the degree of crystallinity. Similarly as in crystals of low-molecular-mass compounds, the possibility of a chemical reaction in an ideal macromolecular crystal is rather limited. Molecules in crystals occupy the available space economically as evidenced by the packing coefficient [19] given as the ratio zV_0/V, where z is the number of molecules, each of inherent volume V_0 located in the basic crystal unit cell of volume V. The packing coefficients for crystals of polyethylene, cellulose and poly(1-butene) are 0.70, 0.81 and 0.63 respectively. As anticipated, the degree of packing of chains in crystals is enhanced by strong intermolecular interactions such as hydrogen bonding, as in cellulose, and is reduced for polymers with bulky side groups. Depending on the packing coefficient, ideal polymer crystals are almost inpenetrable to small molecules of reactants and chemical attack is extraordinarily difficult. Fragmentation reactions induced by radiation are more probable, but even if intramolecular decomposition occurs, for example into two radicals, the impossibility of their diffusional separation brings about their recombination. Therefore, only a small fraction of macroradicals is decomposed into smaller molecules and radicals. However, any defect created in the original crystals by reactions becomes sites of subsequent, faster and more efficient reactions.

In contrast to the solids formed by small molecules, the structure of polymers in the solid state is much more complex. The characteristic forms of intermolecular organization of chains are shown in a simplified way in *Fig. 2.7*. Example *a* illustrates the *solid amorphous state* with interpenetrating individual coils. Macromolecular chains in the ideal amorphous state exhibit only some short-range correlations, but no long-range ordering typical of the crystalline phase is observed. At lower temperatures, the kinetic energy of segmental motion is insufficient to overcome intermolecular forces and the translocation of chain segments ceases; the polymer enters the glassy state and becomes hard and brittle. The volume of the glassy polymer is about 10—20 % higher than that in the crystalline state and the possibility of diffusion of a reactant into "cavities" present in the glassy polymer brings about an enhancement of reactivity relative to the ideal crystal.

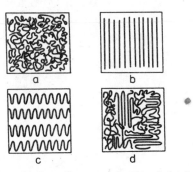

Fig. 2.7. Types of organization of macromolecular chains in the solid state
(*a*) amorphous state with mutually interpenetrating coils; ideal crystal arrangement with (*b*) extended and (*c*) folded chains; (*d*) semicrystalline structure

The accessibility of reaction centres to reactants is markedly increased when the polymer is heated above the glass-transition temperature T_g and is converted into the rubbery amorphous state. The polymer matrix can be likened to a highly viscous medium in this temperature range and the reactivity of functional groups should adjust correspondingly. From this reasoning a dramatic change of the reaction rate should be anticipated at temperatures close to T_g.

A monomolecular isomerization reaction provides a suitable example to illustrate the change in reactivity at T_g. Visually, a very impressive demonstration of this process is the change in colour of spirobenzpyrans in a polymer matrix following irradiation by a UV source [20]

33

The scission of the C—O bond and the isomerization which follows brings about the formation of brightly coloured merocyanines, which can be reconverted thermally to the colourless compounds.

The rate of decolourization depends strongly on T_g regardless of whether the chromophore is "dissolved" in the amorphous polymer matrix or is attached to the chain. The rate constant in both cases is lower than for the analogous reaction in solution. The Arrhenius dependence of the rate constant of the first-order decolourization reaction for poly(isobutyl methacrylate) with 0,63 mol % of randomly placed units of substituted spirobenzpyran is shown in *Fig. 2.8*. It is seen that the discontinuity is located near T_g. The Arrhenius parameters in the regions below and above T_g are quite different; for example the activation energy is 63.2 and 135.2 kJ mol^{-1} below and above T_g, respectively. The question of the sensitivity of isomerization reactions to the change of free volume in an amorphous polymer can be reversed, and these compounds can be used as probes of the structural order in polymers.

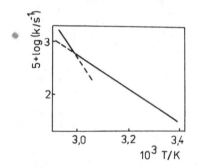

Fig. 2.8. Arrhenius plot of the rate constant of decolourization of the poly(isobutyl methacrylate) derivative with spirobenzopyran units [20]

Returning again to *Fig. 2.7*, the two possibilities of organization of chains in ideal crystals of synthetic and some biological polymers are shown in structures *b* and *c*. The type *c* structure is found in polymer monocrystals with folded chains, formed by crystallization from solution and with a thickness of the

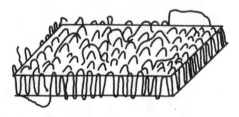

Fig. 2.9. Model of the surface of a lamellar crystal. The chains can re-enter the lamella either in adjacent sites or randomly, thereby forming large loops

microscopic lamella of about 5—20 nm. The macromolecular chains traverse the lamellae back and forth. Structure *b* in *Fig. 2.7* corresponds to crystals with extended chains and the length of the crystallite is correspondingly increased up to, say, 2000 nm.

We have already mentioned the conditions unfavourable to chemical reactions in idealized crystal structures of polymers. In reality, however, polymer crystals involve defects (*Fig. 2.9*) which are more numerous than is common in other crystalline materials. This fact is already evident from a comparison of the density of a crystalline polymer with the value calculated from the unit cell structure. The effective degree of crystallinity may reach only about 0.9 even for single crystals of polyethylene prepared from solution. There are various types of defects in polymer crystals, most of which are connected with the chain folds. The chains may form shorter or longer loops at the crystal surface before re-entry into the lamellae. The chain folds, together with free chain ends and with entire uncrystallized chains, lend an amorphous character to the lamellar surface. This amorphous layer can be of considerable thickness, for example up to 3 nm on each side of the 16 nm thick lamella of a polyethylene crystal. Moreover there are in the amorphous layer segments of chains of polymers with two end-tails crystallized in two *different* lamellae. Surface defects in crystals comprise therefore a major site of chemical reactivity due not only to the accessibility of the chains but also to the stress within them.

The concepts described above are supported by numerous experiments involving chain scission. For example, the hydrolysis of polyethylene terephthalate crystallized from the melt yields oligomers [21] with mean lengths

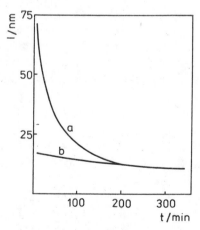

Fig. 2.10. Viscometric determination of the length
(*a*) of poly(ethylene terephthalate) molecules and (*b*) the width of the polymer crystal lamella, both as a function of the hydrolysis time [21]

decreasing with the time of hydrolysis until they reach the width of the lamella (*Fig. 2.10*). The final product, with a strictly homogeneous chain length, is formed from lamellar crystals of equal thickness, while the mass loss during reaction corresponds exactly to the mass fraction of noncrystalline poly(ethylene terephthalate).

Similar results were obtained from the oxidation of lamellar crystals of polyethylene by nitric acid or other oxidants. The thickness of the lamellae of folded crystals of polyethylene was determined from the change in molecular mass after oxidation. The carboxylic acids formed in this way can be transformed into alkanes of equal length, providing a useful method of preparing monodisperse oligomers. This method is applicable for any material crystallizing in lamellae with folded chains. When more drastic treatments are used, such as the low-pressure discharge of gases, the reactions do not stop at the lamellar surface.

Lamellae with a regular organization of chains in folded or extended crystals associate into more complex forms such as spherulites, fibrils, etc. The resulting supermolecular structure and overall macroscopic shape is denoted as the morphology of the crystal. The *semicrystalline structure* shown in *Fig. 2.7* as example *d* is typical for the majority of crystallizable polymers and it is formed by rapid cooling of the melt. This model, termed a fringed micelle, is characterized by the presence of a number of small crystallites and by the large amount of amorphous or quasiamorphous material arising from irregular chain folds, of intercrystallite tie molecules, etc. The degree of crystallinity in semicrystalline polymers can reach figures as low as about 0.8 to 0.4 in polyprophylene, depending on the thermal history of the sample, and even in such typical fibre-forming polymers as polyamides it is not larger than 0.6.

The disordered regions are integral parts of the structure of real "crystalline" polymers and their two-phase character is visualized in the model of fringed micelles. It is obvious that the materials from the two-phase semicrystalline polymers exhibit much higher reactivity compared with perfectly crystallized materials, and the enhancement of reactivity is proportional to the degree of disorder. It is the interface between quasiamorphous and crystalline phases which is critically important as regards reactivity. It should be pointed out, however, that the magnitude of "disorder" in the quasiamorphous phase is hard to define and is influenced by the thermal history of the sample or by the chain orientation when drawing the sample.

The *influence of morphology on reactivity* is particularly apparent for polymers with several structural modification such as cellulose. The accessibility of the reaction centres in semicrystalline polymers to small molecular reagents is specified by the diffusion coefficient D_{sc}. Generally, the coefficient D_{sc} is always lower than the corresponding diffusion coefficient of a reagent in an amorphous polymer D_{am}

$$D_{sc} = D_{am}/\tau\varrho$$

The reduction factor τ accounts for the tortuosity of the diffusional path and ϱ for its impedance. It means that firstly the penetrant must travel further than in the homogeneous amorphous material because of the impermeable crystalline regions. The second factor takes into account the narrowness of some passages in the amorphous regions, which impede or block the penetrant.

The variations in the kinetics of oxidation of samples of polyethylene differing only in morphology, shown in *Fig. 2.11*, can be explained by the varying accessibility of the functional groups. In the first stage of reaction, the surface layers of the powdered monocrystals are oxidized most rapidly since they are more easily penetrated by oxygen than is the film. The higher fraction of the amorphous phase in an unoriented polyethylene sample is evidenced in the later stages of reaction. The fraction of the amorphous phase decreases as a result of the orientation of the film almost to the level found in monocrystals.

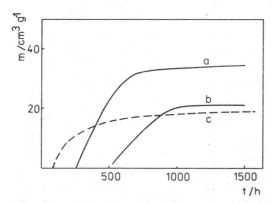

Fig. 2.11. Effect of morphology of linear polyethylene on its oxidation kinetics at 100°C; m — amount of oxygen consumed per gram of polymer
(a) unoriented, (b) oriented film and (c) powdered monocrystals of polyethylene [22]

Conversely, the variation of reactivity with morphology can be used to estimate the amount of disordered material in samples of a given polymer. Obviously, this depends on the diffusional properties of the reactants as well as on their reactivity as it is seen from *Table 2.2*. Here the fraction of disordered material capable of reaction is compared for the four various types of cellulose differing according to the structure of the crystallites [23]. Apart from the large difference in reactivity of the same reagent towards the various types of ce-

Table 2.2   **Percentages of the disordered phase in various celluloses [23]**

Method of determination	cotton	wood pulp	mercerized cotton	regenerated cellulose
X-ray diffraction	27	40	49	65
Density	36	50	64	65
Moisture sorption	42	49	62	77
Deuteration	42	55	59	72
Iodine sorption	13	27	32	52
Acid hydrolysis	10	14	20	28
Alcoholysis	10	15	25	—
Periodate oxidation	8	8	10	20
NO_2 oxidation	23—43	—	—	40—57
Formylation	21	31	35	63

llulose, one should note the variation in percentage of material which has reacted for the same type of cellulose, but with different reagents, e. g. between 8 and 42 % for cotton.

The simplified classification of structural order according to *Fig. 2.7* can also be used for the whole range of biopolymers in rationalizing their reactivity, but it does not include the globular proteins in their natural state. A very ʻght packing of chain segments is typical for the *globular structure*, and is unique ʻor each type of protein. The degree of chain packing in some parts of globular biopolymers is comparable to that in common molecular crystals. Hence it is not surprising that a reactive group localized within the globular structure may be quite inaccessible to a particular reagent. The reactivity of the group can often be manifested only after structural rearrangements when the globular form is broken down into the random coil of the denatured protein featuring "exposed" functional groups. For example, the thiol groups on cysteine are easily oxidized in low-molecular-mass compounds, but not in ovalbumin in its natural conformation; the reaction proceeds only after denaturation of the protein.

G. REACTION KINETICS OF SOLID-STATE POLYMERS

In the investigation of the solid-state reactions, an unexpected deceleration of reaction can frequently be observed [24], but a change in the reaction conditions, such as an increase in temperature can bring about reinitiation of the process (*Fig. 2.12*). This stepwise dependence reflects the kinetic non-equivalence of the reacting compounds originating from differences in their conformational arrangement, from the different structure of their local environment, their distribution in the sample and their mobility. The whole reaction system can be charac-

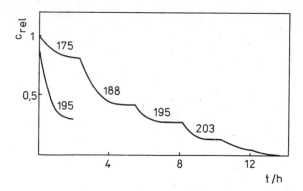

Fig. 2.12. Stepwise decay of poly(methacrylic acid) radicals at various increasing temperatures (K)

terized by a distribution of rate constants for individual, kinetically nonequivalent subsystems; consequently, we can refer to *"polychromatic" kinetics.*

Polychromatic kinetics can be observed in the majority of processes taking place in solid-state polymers such as radical pair recombination after irradiation of the polymer by an ionizing source, solid-phase polymerization, etc. Particularly marked step-like dependences are found with bimolecular reactions below the glass transition temperature when the system solidifies unevenly.

The kinetic nonequivalence of reagents in the solid state is one of the origins of the compensation effect, which is usually expressed as a correlation between the experimentally determined preexponential factors A and the activation energies E

$$\ln A = a + bE$$

where a and b are constants. This relationship is not unique to reactions in the solid state. Here, however, it may be accentuated by an improper selection of the model for determination of the rate constant. When instead of the correct model

$$g(\alpha) = kt$$

where k is the rate constant, t is time and α the extent of reaction, the wrong model $G(\alpha)$ is used for the description of reaction, then a simple relationship exists between the correct and apparent activation energies E_{cor} and E_{app}, respectively [25]

$$E_{app} = E_{cor} - RT \ln [G(\alpha)/g(\alpha)]$$

In solid-state reactions the selection of a "correct" model represents only an approximation to the real reaction mechanism and confusion of $g(\alpha)$ and $G(\alpha)$ is very frequent.

39

Let us consider the variations of the parameters of the autocatalytic reaction

$$X \xrightarrow{k_1} Y$$

$$X + Y \xrightarrow{k_2} 2Y$$

correctly modelled by the function $g(\alpha) = \ln[\alpha/(1 - \alpha)]$ but treated under non-isothermal conditions by the first-order reaction scheme $G(\alpha) = \ln(\alpha)$.

The temperatures dependences of the rate constants are expressed by the equations

$$k_1 = A_1 \exp(-E_1/RT) \qquad k_2 = A_2 \exp(-E_2/RT)$$

where A_1, A_2, are pre-exponential factors and E_1, E_2 are activation energies. The calculated dependence of the instantaneous concentration of component X on temperature is shown in *Fig. 2.13*. The group of four parameters A_1, A_2, E_1 and

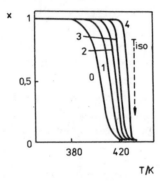

Fig. 2.13. Variation of the relative concentration x of compounds X with temperature on the programmed heating of the system described by equations $X \rightarrow Y$ and $X + Y \rightarrow 2Y$ for the following parameters

initial temperature $T = 293$ K, heating rate 5 K min^{-1}, $A_2 = 1 \times 10^9$ s^{-1}, $E_2 = 83.7$ kJ mol^{-1}, $E_1 = 146.5$ kJ mol^{-1} and variable A_1 (in s^{-1}): 1×10^{16} (0), 1×10^{15} (1), 1×10^{14} (2), 1×10^{13} (3) and 1×10^{11} (4)

E_2 corresponding to the two rate constants is reduced to the pair A' and E' in the single rate constant model, $G(\alpha)$, which fits satisfactorily to the above of concentration curve. For example, assuming $E_1 = 146$ kJ mol^{-1}, $E_2 = 83.7$ kJ mol^{-1}, $A_2 = 1 \times 10^8$ s^{-1} and A_1 lies in the range between 1×10^{11} — 1×10^{16} s^{-1}, the activation energy E' varies in the range $152 - 721$ kJ mol^{-1} and $\ln A'$ in the range $31.1 - 172.5$. The variations in one parameter of the original model (A_1 in our case) can be related to a change in the physical state of the polymer matrix and in the distribution of the rate constants associated with the kinetically nonequivalent subsystems. These changes in A_1 produce, following the formal treatment, a whole series of pairs of A' and E' with mutually compensating values (*Fig. 2.14*). Thus unrealistically high or low values could be assigned to the activation energy E'.

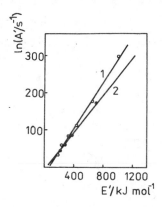

Fig. 2.14. Compensation of the values of ln A' and E' found for the autocatalytic system with linear heating from the first-order reaction scheme using various parameters for A_1 in the range $1 \times 10^{11} - 1 \times 10^{16}$ s^{-1} and for $E_1 = 146.5$ kJ mol^{-1}, $E_2 = 83.7$ kJ mol^{-1} and $A_2 = 1 \times 10^9$ s^{-1} (line 1) and 1×10^8 s^{-1} (line 2)

An isokinetic temperature T_i can be specified ($T_i = 444$ K in *Fig. 2.13*) at which the physical barriers to the chemical conversion of compound X disappear. It is evidently the temperature of the phase transition (melting) in the system.

The diffusion-controlled bimolecular reactions of a polymeric reagent X and a low-molecular-mass compound Y, can be divided into three important stages:
1. macrodiffusional, during which the concentration gradients of compound Y equilibrate in the system,
2. microdiffusional, in which the rate of the process is determined by the rate of formation of the pairs $X...Y$,
3. the kinetic stage of the actual reaction $X...Y$ in the pre-formed pair.

Apart from physical diffusion, *chemical diffusion* is important in the reaction of radicals in polymers as can best be illustrated by the isoergonic transfer of a free spin within a polyethylene chain

$$-CH_2\dot{C}HCH_2- \; + \; -CH_2CH_2CH_2- \quad \rightarrow \quad -CH_2CH_2CH_2- \; + \; -CH_2\dot{C}HCH_2$$

or in the repetition of the cycle of reactions during the oxidation,

$$PO_2^{\cdot} + PH \quad \rightarrow \quad P^{\cdot} + POOH$$

$$P^{\cdot} + O_2 \quad \rightarrow \quad PO_2^{\cdot}$$

41

This migration mechanism is termed a *"relay transfer"* [24] and can be described by the effective diffusion coefficient D_{eff}

$$D_{eff} = k_p \lambda [RH]/6$$

where λ is the mean distance of transport of the radical site, k_p is the rate constant of the rate-determining step in the relay and [RH] is the concentration of monomer units. A "unit step" of the general order of several tenths of a nanometer is indicated by estimation of D_{eff} for the relay reaction. Therefore, the mechanism of the relay reaction needs to be supplemented by the diffusion of the particles, especially of the small radicals, which mediate the migration of the reaction site. There are several essentially equivalent ideas about the formation of low-molecular-mass radicals from macroradicals. The possibility of fragmentation of macroalkylperoxyl radicals through the six-membered cyclic transition-state

with the formation of a carbonyl group, alkene and hydroxyl radical [26] is ruled out by the high reactivity of hydroxyl radicals which does not allow their transport far from their place of generation.

A more probable transport of free valence is that mediated by $HO_2^{\cdot}$ radicals, which are less reactive as regards abstraction of hydrogen from the polymer chain and which can be formed from peroxyl macroradicals

Low-molecular-mass radicals can also arise from alkoxyl or alkyl radicals by β-scission

The idea of migration of free valence is attractive and supported by several arguments. One has to bear in mind however, that postulation of the original but unproven steps to substantiate the formation of $HO^{\cdot}$ and $HO_2^{\cdot}$ is at least equivalent to a mechanism for their production via the mono- and bimolecular decomposition of macromolecular hydroperoxides.

Some aspects of solid-state kinetics may be modelled by the elementary steps of transfer reactions of free radicals. For example, the elimination of an H-atom from an sp^3-hybridized carbon atom in a chain is associated with the subsequent rehybridization into sp^2- carbon with a concomitant change of the valence angles from 109° into 120° and the appropriate change of bond lengths. In the solid state, the process of rehybridization and structural relaxation may not be synchronous with the chemical step, thereby increasing the activation energy of a given elementary step. An alternative model assumes a suitable orientation of the reactants as a prerequisite for the occurence of reaction.

Let us consider now how the effect of local concentration, occasioned for example by the preferential inclusion of the reactant into domains in the polymer matrix, is reflected in solid-state reactions. When a compound M reacts according to the first-order reaction scheme

$$M \xrightarrow{k_1} N$$

and assuming that n_M mole is distributed homogeneously, the reaction rate is given as follows

$$v = k_1 n_M / V$$

However, when in a system with overall volume V, i heterogeneous regions of compound M are formed with a volume V_i and with an unchanged reactivity of M for a given type of reaction, and when the redistribution of M into the overall volume V is prevented, the reaction rate is given by

$$v' = k_1 \Sigma (n_{Mi} / V_i)$$

where n_{Mi} is the amount of substance M in microheterogeneous zones with volume V_i. Since for each V_i the relation $V_i \ll V$ applies, it is evident that $\Sigma n_{Mi} / V_i \gg n_M / V = \Sigma n_{Mi} / V$ and $v' \gg v$ i.e. the reaction rate in a heterogeneous system can be substantially larger than in a homogeneous system. The overall acceleration of reaction, caused obviously by the accumulation of material in certain zones of the total reaction volume, can, on the other hand, be compensated by the reduction of the intrinsic reactivity of the substance M in these zones. At all events, however, the variation in concentration of reactants inside the polymer volume and the re-distribution of concentration heterogeneities under the particular conditions of reaction are the sources of a frequently observed characteristic solid-state reactions, namely, of an acceleration of the process after the intial induction period.

H. REACTIONS CATALYZED BY POLYMERS

Chemical reactions of macromolecules also incorporate processes in which the polymer acts as a catalyst. There are diverse roles for polymers in catalytic processes, from acting as the relatively passive support for attachment of catalytically active groups to the highly effective and selective enzymes. Polymer catalysts can affect the reaction rates of low-molecular-mass compounds as well as of macromolecular reagents.

In reactions of low-molecular compounds, a polymer catalyst may either simply provide adsorption sites on the macromolecule or it may include the catalytically active groups which are linked to the chain. The first category involves mainly the reactions of low-molecular-mass ionic substances with the accelerating or retarding effect of polyelectrolytes possessing no catalytically specific groups. As noted during discussion of the effect of the medium on polyionic reactions, the catalytic power of polyelectrolytes stems from their ability to concentrate or to repel ionic reagents. Addition of polyelectrolyte may sometimes bring about dramatic effects as in the case of reaction

$$[Co(NH_3)_5Cl]^{2+} + Hg^{2+} + H_2O \quad \rightarrow \quad [Co(NH_3)_5H_2O]^{3+} + HgCl^+$$

in water, where the presence of poly(vinyl sulphonate) ($-CH_2CHSO_3-)_n$ accelerates the reaction by a factor of 1.76×10^5 [27]. The catalytic action is reduced by the addition of a simple electrolyte which diminishes the electrostatic potential in the domain of the macromolecule.

In contrast, the reactions of ionic compounds with opposite charges may be retarded by addition of a macrocation or macroanion. For example, Wohler's synthesis of urea

$$NH_4^+ + OCN^- \quad \rightarrow \quad (NH_2)_2CO$$

is retarded either by poly(acrylic acid) or by the copolymer of diethyldiallylammonium chloride and sulphur (IV) oxide [28]. When the reaction does not take place in water but for example in a mixture of water and an organic solvent, the reaction rate is affected not only by the coulombic interaction of the ions but also by solvation effects such as the preferential hydration of the polyelectrolyte.

The acceleration or retardation of reaction by the addition of polymer is not limited to ionic systems. A nice example of the reactions of nonelectrolytes in organic solvents is provided by the reaction of dansyl chloride (5-dimethylamino-1-naphthalene-sulphonyl chloride) with a primary amine in chloroform, i.e. the same model system as discussed before concerning interchain reactions. The second-order rate constant k_2 of this reaction decreases with the addition of toluene to the chloroform solvent [29]. This decrease is considerably attenuated when polystyrene with a molecular mass of up to 17 500 is used as a cosolvent. However, the same polymer cosolvent enhances the reaction of dansyl chloride with a primary amine end-linked to polyethylene. The variation of the reaction rate of electrolytes and nonelectrolytes on the addition of macromolecules can be described generally by the Brönsted equation

$$k_2/k_2^0 = \gamma_A \gamma_B/\gamma_X$$

where k_2^0 is the rate constant of reaction in an ideal solution and γ_A, γ_B, γ_X are the activity coefficients of the reactants and transition state X, respectively. The activity coefficients can be estimated from the solution theories of polymer electrolytes or nonelectrolytes.

The second much more important category of polymer catalysts involves linkage of the chains to catalytically active groups. The importance of this class is due in part to its relevance to enzymatic systems. One of the long-term aims of investigation in this area is the understanding of the nature of enzymatic action in the catalysis of biochemical reactions. The role of the diverse types of interaction in the mechanism of catalysis is being elucidated gradually by means of suitable model systems. The successful preparation of enzyme-like systems from macromolecules of non-biological origin has been carried out. It is evident that the synthetic polymer catalysts (also termed "synzymes") do not possess the unique tertiary structure typical of the globular enzymes, yet they are comparable to or may even surpass enzymes in some catalytic properties. The synthetic polymer catalysts usually exhibit not only a higher catalytic action than the corresponding small-molecule catalyst but may show some specificity towards substrate, competitive inhibition by substances resembling the reactive substrate, saturation effects and similar phenomena typical for enzyme kinetics.

The enhanced efficiency of a polymer catalyst originates from numerous factors; the concerted action of several catalytic groups on the reactive centre is one of them. For example, imidazole has pronounced nucleophilic properties which function in the catalysis of hydrolytic reactions. It turns out, however, that poly(2-vinyl imidazole) [30]

is a four-fold more effective catalyst than imidazole in the hydrolysis of the neutral ester of 4-nitrophenyl acetate at high pH according to the scheme

This enhanced effect is ascribed to multiple cooperative catalysis by anionic and neutral imidazole groups in the polymer chain. (This reaction is often used in enzymology for the investigation and testing of hydrolytic enzymes.)

When an amine-water mixture is used as solvent instead of water, aminolysis of the ester occurs. The investigation of the rate of decomposition of two acylnitrophenyl esters by amines of various molecular mass [31] shows (*Table 2.3*) that the reaction rate increases moderately with the polyfunctiona-

Table 2.3   **First-order rate constants ($k_1/100$) for the aminolysis of 4-nitrophenyl esters [31] (in min^{-1})**

Amines	4-nitrophenyl acetate	4-nitrophenyl laurate
Propylamine	0.98	0.053
PEI 600[a]	3.60	0.11
PEI 1 800[a]	4.38	0.11
PEI 60 000[a]	4.60	0.17
PEI 600 — L 10 %[a, b]	15.2	698

a) the number shows the molecular mass of polyethyleneimine
b) PEI with of 10% nitrogen atoms acylated by lauroyl groups

lity of the amine but decreases with lengthening of the acyl group from two to twelve carbon atoms. A striking rate enhancement (ten thousand-fold) is achieved when a modified polyethyleneimine (PEI) is used as a catalyst in which about 10% of the nitrogen atoms are acylated by lauroyl groups. The above observation illustrates that hydrophobic interactions between sufficiently long non-polar chains in a catalyst, and a substrate, are more important that the effects of *multifunctional catalysis* or of *electrostatic* and *solvation interactions*. In general, noncovalent *interactions* of *hydrophobic* or other types assist in the fixation of the substrate on the catalyst and in the formation of an equilibrium complex. The association may dramatically enhance the effectiveness of the catalysis when the reactive group of the substrate and the functional groups of the catalyst are juxtaposed and properly oriented.

An even higher catalytic efficiency of PEI was achieved when nitrogen groups, in addition to being reacted with dodecyl iodide, were alkylated with chloromethylimidazole and in this way the methyleneimidazole group

$$-CH_2-\underset{\underset{\underset{CH}{HN}}{|}}{C}=\!\!=\!\!CH$$

has been introduced into PEI. Using PEI with 10 % of lauroyl and 15 % of methyleneimidazole groups, a rate enhancement by twelve orders of magnitude was observed in the hydrolysis of 2-hydroxy-5-nitrophenyl sulphate, one of the largest values accomplished with synthetic polymer catalyts [31]. Interestingly, PEI modified in this manner is a more effective catalyst of the above reaction than the natural enzyme arylsulphatase.

The first-order rate constants (*Table 2.3*) can provide values for the so-called "catalytic constant" k_{eff} [31] and the efficiency of various catalysts in the hydrolysis of nitrophenyl esters can be compared. The catalytic effect of modified PEI relative to imidazole is large, being comparable with chymotrypsin (*Table 2.4*).

Table 2.4   **The relative effectiveness of various catalysts in the hydrolysis of 4-nitorphenyl acetate [31]**

Catalyst	k_{eff} /dm^{3}mol^{-1}min^{-1}
Imidazole	10
α-Chymotrypsin	10 000
PEI 600 — L (10 %) Im (15 %)[a)]	2 700

a) Polyethyleneimine with molar mass 600 g mol^{-1} with 10 and 15 % (mol) of nitrogen atoms substituded by lauroyl and methylene imidazole groups, respectively.

The efficiency and selectivity of synthetic and biological polymer catalysts is mainly affected by steric conditions at the catalytic site and especially by the mutual fitting of the shapes of catalyst and reactants. The identical tertiary structure of enzymes of a given type secures the optimal spatial configuration of all the groups active in the catalytic act. In order to achieve this optimal orientation of groups, the active region of the enzyme tends to be located in a slit or groove formed in the overall globular structure. Evidently, the shape of the reactant (substrate) molecule should conform to that of the enzyme slit as illustrated in the popular visualisation of the enzymatic action by the lock-key analogy.

In an attempt to simulate the enzyme-like spatial configuration of catalytic groups in synthetic enzymes, one needs either to start from suitable molecules with a cavity in their structure and subsequently modify them, or to build up by a tailor-made synthesis macromolecular systems containing both the cavity and the required spatial configuration of catalytic groups. In the first approach, molecules forming inclusion complexes are used, such as crown ethers and cyclodextrins (cyclic oligomers of glucose with 6—8 units). The catalytic activity of cyclodextrin is promoted by hydrophobic substituents on one side of the cavity, and by linkages to imidazole groups, etc.

As an example of the second approach, we cite the means of achieving a cavity in a synthetic polymer catalyst by template polymerization [32]. The active groups which should ultimately be located at the catalytic centre are linked to an appropriate template molecule in the form of polymerizable vinyl derivatives (*Fig. 2.15*). The subsequent copolymerization and crosslinking provide the polymer with a fixed chain structure. On removing the template, "free" cavity is produced with required shape and with appropriate positions of the active groups.

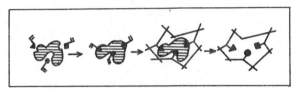

Fig. 2.15. Formation of the active centre of a synthetic polymer catalyst by template polymerization [32]

This approach aims not so much to achieve rate enhancement but to increase the specificity of the catalytic action. A synthetic catalyst prepared by template polymerization can differentiate the very fine structural variants such as the L and D isomers (enantiomers) of complex saccharides [32] with the same selectivity as biological receptors.

The catalytic effect is not observed when a sufficiently close approach of the substrate to the active centre is hindered, for example, by steric effects. For example, the hydrolysis of the copolymer of acrylamide and monomers with L-phenylalanine-4-nitroanilide residues linked by spacers of variable length [17] is revealing

$$-CH_2CH-$$
$$CONH-(CH_2)_n-CONHCHCONH-\bigcirc-NO_2$$
$$CH_2$$
$$\bigcirc$$

$n = 1, 3, 5$

It is found that the rate of hydrolysis of the 4-nitroanilide group in these copolymers by the action of chymotrypsin is comparable to the reaction rate for a low-molecular-mass substrate only if the hydrolytically-sensitive bond is separated by eleven covalent bonds from the chain backbone. Shortening of the side-chain spacer to nine or seven bonds substantially reduces the activity of the chymotrypsin. This retardation is apparently connected with steric barriers to the approach of the side-chain groups to the catalytic centre in the enzyme.

The catalytic process in the enzyme groove depends on cooperation between the so-called binding sites and the intrinsic catalytic centre. The binding sites fix the substrate inside the groove by noncovalent interactions, and the stabilization of the resulting complex depends on the number of interactions. From this point of view, the longer, or at least oligomeric substrates, are favoured over small ones and their reactions are subject to greater acceleration. Noncovalent interactions may be hydrophobic (as in the case cited of polyethyleneimine), hydrogen bonds, Coulomb interactions between charges on the enzyme and substrate, and of other types. In general, the stronger the association of substrate and enzyme through the binding sites, the more effective is the enhancement of a reaction by the catalytic process.

The textbook example of cooperation between the binding sites and the catalytic centre is lysozyme. Lysozyme was the first enzyme for which the three-dimensional structure was determined and, at the same time, the mechanism of its action was explained from the crystallographic data [33]. Lysozyme is found in bodily secretions such as tears and is especially abundant in egg-white. The lysozyme chain contains 129 amino acid residues and the resulting globular shape with a groove on one side is stabilized by four disulphide bonds. Lysozyme cleaves polysaccharides, forming a protective wall in bacterial cells, by hydrolysis of glycosidic bonds between pyranose units. It has a large active centre conventionally divided into parts A to F, each of which binds one saccharide unit of a macromolecular substrate. The actual catalytic act takes place at site D, the other sites serving as binding sites for stabilization of the enzyme-substrate complex. Two carboxyl groups in the vicinity of site D belong to glutamic and asparagic acids, located as the 35th and 52nd residues respectively, in the primary structure of the lysozyme chain (Glu-35 and Asp-52). In the catalytic process, the undissociated carboxyl group on Glu-35 transfers its proton to a glycosidic oxygen, and the scission of the C—O bond between the sugar units takes place with formation of a carbonium ion in ring D. The proton transfer is facilitated by the ionizable carboxyl group on Asp-52 which electrostatically stabilizes the emerging cation

The hydrolysis terminates when the carbonium ion reacts with HO^- derived from a water molecule and the remaining proton is transferred to Glu-35. In this way the enzyme, which is modified during the catalytic process, regenerates its original state. Finally, the products of hydrolysis have to diffuse away from the active site.

Of course, the detailed mechanism of catalysis is different for each enzyme but common elements remain, such as the presence of binding sites, the formation of an enzyme-substrate complex, transient transformation of the enzyme during the act, etc. Much research effort is currently concentrated on the question of the origin of the high catalytic efficiency of enzymes. In comparison with simple catalysts, the enzymes are by some 8—9 orders of magnitude more powerful in their rate enhancement of a reference reaction. It is widely assumed that this striking enhancement is brought about by the summation of numerous diverse effects, which individually are much smaller. We have already mentioned some of these factors such as the mutual steric fit, the hydrophobic and electrostatic interactions and the polyfunctionality of the catalyst. The activation energy of reaction can also be reduced by the local microenvironment (the macromolecular chain) at the catalytic centre, by fixation of the optimal conformation of the reactants in the transition-state, by securing the optimal angle of approach of the reactants (orbital steering), etc. It is very difficult to separate these individual contributions in real enzymatic systems. Detailed elucidation of the mechanism of enzymatic activity requires full knowledge of the three dimensional structure of the enzyme, supplemented by kinetic data such as the reactivity of the substrate and its modified analogues. An understanding of enzymatic catalysis is relevant not only to the problem of catalysis in general but especially to the design and synthesis of the enzyme analogues based on non-protein polymers.

I. DEFECT CENTRES IN MACROMOLECULES

The treatment of the polymer effect in the reactions of macromolecular compounds would be incomplete without mentioning the role of the "anomalous" or "defective" structural units which are always incorporated into a chain of "normal" mers. The presence of the defective structural mers is a consequence of the formation of secondary by-products. In low-molecular-mass reactions, the by-products are usually removed by some separation technique and, after purification, molecules of uniform structure are obtained. The situation is different in polymer synthesis; here the products of side-reactions are frequently integrated into the polymer chains as "anomalous" structural units.

The number of anomalous mers in a chain depends on the difference between the Gibbs free energies of formation of the regular and anomalous mers, on the type of reactive intermediate in the polymerization and on the reaction conditions. Mers with a structure similar to, or isomeric with the regular mers are mainly built into the chains. Consequently an irregular atactic chain is usually formed on radical polymerization of vinyl monomers, representing the statistical copolymer of the two stereoisomeric forms. The regular chain of ordered mers found in iso- and syndiotactic polymers growth mainly by stereoselective opening of the double bond.

The occurrence of a major deviation in a regular sequence of mers is rare, yet its influence on the chemical and biological activity of macromolecules is much higher. The marked influence of a small concentration of defects on the properties is explained by the macromolecular character of the reagent and by the subsequent chain reactions. For example, the molecular mass of a macromolecule is reduced, and typical polymer properties are lost, when the chain is broken at only a few points which originated as defect mers. Structural anomalies are also a potential source of crosslinks and they also affect considerably the properties of macromolecular materials. Generally, minimal amounts of defect isomers are incorporated into the growing chain during anionic polymerization of alkenes, although more appear during polymerization of vinyl monomers by radical or cationic mechanisms. The influence of temperature is very pronounced, for example, the radical polymerization of styrene at 130°C introduces seven defects of quinonemethide type

$$A—[—CH_2CH]_x—CH_2CH=⟨⟩=CH\ [CH_2CH]_y—B$$

per macromolecule as against a figure of only two at 30°C.

Defect centres may also include unwanted components originally present in an unsufficiently purified monomer or introduced adventitiously from the ambient atmosphere. In this way, the thermal stability of polymers is frequently reduced due to reactive labile peroxidic species formed by radical copolymerization of monomers when oxygen is incompletely removed during polymerization.

Irregularities in macromolecular growth are also apparent in some other polyreactions. The polycondensation of ethylene glycol and terephthalic acid yields not only the structural units

$$[—OCC_6H_4COO(CH_2)_2O—]_n$$

but also in small amounts, the ether-linkage unit

$$—OCC_6H_4COO(CH_2)_2O(CH_2)_2OCOC_6H_4COO—$$

Any breakdown of regularity in the sequence of structural units diminishes the crystallinity of the product, and thus the reactivity of the defect unit increases further. The synthesis of macromolecules in vivo is more regular, but even in this circumstance macromolecules are not produced without defects. Natural rubber contains 97—98 % of *cis*-1,4-isoprene mers and the rest consists of various structural units.

Anomalous units are incorporated also in the functionally more elaborate and structurally more complex biopolymers. During the biosynthesis of these molecules, the healthy organism is usually, but not always able to repair the defect. Some permanent defect can be imprinted into the DNA and bring about the mutation or even extinction of an organism.

The end-groups constitute a special kind of defect in macromolecules since they must differ from the interior units in the chain at least from valency considerations. The type of end group depends on the nature of the polyreaction and the reaction conditions. Considering the chain length, the end groups have negligible effect on the polymer properties of the polymer and they are not represented in the usual macromolecular formulae. As ever, there are exceptions to the rule since the end groups do sometimes affect the properties of polymers such as their thermal stability. This happens especially in those cases when the end groups can initiate the chain reaction and thus the concentration of end groups becomes multiplied by the number of induced reactions. Anomalous units are also formed in macromolecules by the process of ageing, and of course unused initiators and catalysts, which may be difficult to remove, contribute to this process.

These various examples indicate that the infrequent but reactive anomalous mers could be important in the chemical reactions of polymers [34]. This possibility has an analogy in the striking influence of the crystal lattice defects on the strength of solid materials. In contrast to solid-state defects, macromolecular defects have been investigated to a much lesser extent since few suitable analytical methods are available for the determination of the detailed structure of anomalous mers present in minimal concentration. Therefore, the influence of such irregular units is frequently only surmised as the best rationalization of observed facts on the basis of experience gained from model compounds.

References

1. Macromolecules, an Introduction to Polymer Science. BOVEY, F. A., WINSLOV, F. H., (Editors), Academic Press, New York 1979.
2. MORAWETZ, H. in: Chemical Reactions of Polymers. FETTES, E., (Editor), Mir, Moscow 1967, Vol. 1, p. 16—46 (translation into Russian).
3. PLATE, N. A.: Problems of Polymer Modification and Reactivity of Functional Groups in Macromolecules. Pure Appl. Chem., *46*, 49—59, 1976.

4. PLATE, N. A., NOA, O. V., STROGANOV, L. B.: Some Problems of Theory of the Polymer-analogous and Intramolecular Reactions of Macromolecules. Vyskomol. Soed., A 25, 2243—2266 1983.
5. MORAVETZ, H.: Macromolecules in Solution, Wiley, New York 1965.
6. WINNIK, M. A.: Cyclization and the Conformation of Hydrocarbon Chains. Chem. Rev., 81, 491—524, 1981; MITA, I., HORIE, K.: Diffusion-Controlled Reactions in Polymer Systems. J. Macromol. Sci.-Revs. C27, 91—169 (1987).
7. SISIDO, M., MITAMURA, T., IMANISHI, Y., HIGASHIMURA, T.: Intrachain Reaction of a Pair of Reactive Groups Attached to Polymer Ends. I. Intramolecularly Catalyzed Hydrolysis of a Terminal p-Nitrophenyl Ester Group by a Terminal Pyridyl Group on Polysarcosine Chain. Macromolecules, 9, 316—319, 1976.
8. SHIMADA, K., SWARC, M.: Flexibility of Molecular Chains Studied by Electron Spin Resonance Technique. J. Amer. Chem. Soc., 97, 3313—3321, 1975.
9. BLACK, P. E., WORSFOLD, D. J.: Excluded Volume Effect on Polymer Reaction Rates. J. Polym. Sci. Polym. Chem. Ed., 19, 1841—1846, 1981.
10. OKAMOTO, A., TOYOSHIMA, K., MITA, I.: Kinetic Study on Reactions between Polymer Chain-Ends II. Reactions between Chlorosulphonyl-Ended Polyoxyethylene followed by Fluorometry. Eur. Polym. J., 19, 341—346, 1983.
11. PAPISOV, I. M., LITMANOVICH, A. A., BOLYACHEVSKAYA, K. I., MAKAROV, S. V., BARANOVSKY, V. Yu., KAZARIN, L. A.: Intra and Interchain Reactions in Polycomplex Composites. Preprints, IUPAC Symp. Macromol. Chem. Bucharest, Sept. 5—9, 1983, Sect. VI., p. 96—99.
12. CHIU, W. Y., CARRAT, G. M., SOONG, D. S.: A Computer Model for the Gell Effect in Free-Radical Polymerization. Macromolecules, 16, 348—357, 1983.
13. HORIE, K., MITA, I., KAMBE, H.: Fast Reaction and Micro-Brownian Motion of Flexible Polymer Molecules in Solution. Polymer J., 4, 341—349, 1973.
14. JELINEK, H. H. G.: Diffusion and Random Chain Scission. Polymer J.,4, 489—494, 1973.
15. EPTON, R.: Chemical and Biochemical Reactions in Gel Phase, in: Reactions on Polymers. MOORE, J. A., (Editor), D. Reidel, Dordrecht 1973, p. 286—314.
16. SHERRINGTON, D. C.: The Effect of Polymer Structure on the Reactivity of Bound Functional Groups. Nouv. J. Chim., 6, 661—668, 1982.
17. MORAWETZ, H.: Some Recent Studies on Polymer Reactivity. J. Macromol. Sci.-Chem. A 13, 311—320, 1979.
18. LABSKÝ, J., MIKEŠ, F., PILAŘ, J.: Functional Group Behaviour as a Function of Length and Structure of the Spacer Linking them to a Soluble Polymer, (in Czech). Chem. Listy, 77, 1039—1063, 1983.
19. WUNDERLICH, B.: Macromolecular Physics. Vol. I., Academic Press, New York 1973.
20. SMETS, G.: Chemical Reactions of Macromolecules in the Solid State, in: Reactions on Polymers. MOORE, J. A., (Editor), D. Reidel, Dordrecht 1973, p. 371—392.
21. WUNDERLICH, B.: Chemical Reactions Involving the Backbone Chain of Macromolecular Crystals, in: Reaction on Polymers. MOORE, J. A., (Editor), D. Reidel, Dordrecht 1973, p. 395—409.
22. SEGAL, L.: Influence of Morphology on Reactivity, in: Chemical Reactions of Polymers, FETTES, E., (Editor), Mir, Moscow 1967, Vol. 1, p. 46—74 (Translation into Russian).
23. ROWLAND, S. P.: Solid-Liquid Interactions: Inter- and Intra-Crystalline Reactions in Cellulose Fibers, in: Applied Fibre Science. HAPPEY, F., (Editor), Academic Press London 1979, Vol. 2, p. 205—237.
24. EMANUEL, N. M., BUCHACHENKO, A. L.: Chemical Physics of Ageing and Stabilization of Polymers (in Russian). Nauka, Moscow 1982.
25. ŠESTÁK, J.: Thermophysical Properties of Solids. Elsevier Amsterdam, 1984.

53

26. MARCHAL, J.: Oxidative Degradation of Polymers and Organic Compounds in Unimolecular Decomposition of Peroxy Radicals; Mechanism of Pyrolysis, Oxidation and Burning of Organic Materials. NBS Special Publications 357, 85, 1972.
27. MORAWETZ, H.: Chemical Reaction Rates Reflecting Physical Properties of Polymer Solutions. Accounts Chem. Res., *3*, 354—360, 1970.
28. ISE, N.: Polymers with Catalytic Activity, in: Specialty Polymers. ISE, N., TABUSI, I., (Editors), (Translated into Russian), Mir, Moscow 1983, p. 62–92.
29. OKAMOTO, A., HAYASHI, A., MITA, I.: Effect of Polymer as a Cosolvent on Chemical Reactions in Solution. Europ. Polym. J., *19*, 405—408, 1983.
30. OVERBERGER, C. G., SMITH, T. W.: Catalysis by Polymers, in: Reactions on Polymers. MOORE, J. A., (Editor), D. Reidel, Dordrecht 1973, p. 1—24.
31. KLOTZ, I. M.: Synzymes: Synthetic Polymers with Enzyme-like Catalytic Activities, in: Molecular Movements and Chemical Reactivity as conditioned by Membranes, Enzymes and other Macromolecules. LEFEVER, R., GOLDBETER, A., (Editors) Wiley, New York 1978, p. 109—160.
32. WULFF, G.: Selective Binding to Polymers via Covalent Bonds. The Construction of Chiral Cavities as Specific Receptor Sites. Pure Appl. Chem., *54*, 2093, 1982.
33. PHILLIPS, D. C.: The Three-Dimensional Structure of an Enzyme Molecule, in: Biophysical Chemistry, Readings from Scientific American. BLOOMFIELD, V. A., HARRINGTON, R. E., (Editors), W. H. Freeman and Co., San Francisco 1975, p. 141—154.
34. KORSHAK, V. V.: The Chemical Defects in Macromolecules of Heterochain Polymers. Usp. Khim., *42*, 695—742, 1973.

III. MODIFICATION OF STRUCTURAL UNITS IN THE POLYMER CHAIN

The synthesis of a new polymer may be accomplished by polymerization of identical or structurally different monomers or by modification of an existing polymer. In the early days of macromolecular chemistry, this latter route, which will be now the main subject of discussion, was used particularly for modification of the properties of natural polymers. Since direct synthesis of macromolecules of a given structure appeared not always to offer the best route even when possible, attention was gradually focused on improved modifications of synthetic polymers. The synthesis of copolymers with a sequence of mers different from that obtained on direct copolymerization, the synthesis of polymers derived either from unstable monomers or from monomers which do not polymerize, are only a few examples where a strategy of modification of an established macromolecular chain may replace that of direct polymerization.

The chemical transformations of repeating structural units of macromolecules include, of course, the same group of reactions established in the organic chemistry of low molecular-mass compounds. (Here, we confine ourselves to polymer-analogous reactions.) From the large body of knowledge available we shall quote mainly those processes of modification which have found practical application. Also of interest, of course, are processes which alter the functional properties of the macromolecular system or feature in the synthesis of speciality polymers.

Since the reactivity of macromolecular compounds depends critically on the chemical structure of their functional group, it may vary considerably for different polymers under comparable conditions. Just as with low-molecular alkanes, the chemical bonds in saturated hydrocarbon polymers are relatively stable, and reduced reactivity to common reagents is observed also for polymers where the hydrogen atoms are completely substituted by fluorine, as in perfluorocarbon polymers. On the other hand, the presence of multiple carbon-carbon bonds or heteroatoms in the structural units of a polymer chain leads ultimately to an increase in reactivity. The heteroatoms may be an integral part

55

either of the side group or the main chain of the hydrocarbon polymer. While increased reactivity of the side groups widens the spectrum of modification reactions of a given polymer, the effect of heteroatoms in the main chain, where they may initiate the scission of a macromolecule to low-molecular products, is more ambiguous.

A. HYDROCARBON POLYMERS

Chemical transformation of the structural units of a saturated hydrocarbon polymer requires treatment with reactive compounds such as halogens. Of these, fluorine reacts most vigorously. Since very strong fluorine-hydrogen and fluorine-carbon bonds are formed, fluorination is highly exothermic, and to avoid explosion it is usually carried out with fluorine diluted by nitrogen or some other inert gas. Fluorination takes place at the surface layers of a solid polymer; however part of the fluorinated polymer simultaneously decomposes to tetrafluoromethane. In comparison with direct synthesis from fluorocarbon monomers, the modification of polymers with fluorine has not found practical application.

The chlorination of polyethylene and poly(vinyl chloride) [1] is the most frequently-used halogenation reaction. Polymers are chlorinated either in a dispersion in water or when dissolved in tetrachlormethane. The distribution of chlorine atoms along the macromolecular chain depends on the experimental conditions. In heterogeneous chlorination, the chlorine atoms are bound to the surface layers of the polymer particles and to the amorphous chain segments which are readily accessible, while the products of homogeneous chlorination in solution are determined by the chemical reactivity of the substrate. From a homopolymer, random copolymers are formed in a homogeneous process whereas the products formed in the heterogeneous process assume the structure of a block copolymer. Since it was found that the deuterium content measured during the chlorination of α-deuterated poly(vinyl chloride) [2] does not change with increasing amounts of bound chlorine, it may be assumed that the primary attack of chlorine is focused on the CH_2 groups, resulting in one chlorine atom being bound to one methylene carbon atom only.

The polar effect of pre-existing chlorine atoms on the subsequent course of chlorination is evidenced by the fact that for each 90 mol % of reacted methylene groups of poly(vinyl chloride) only 2.6 mol % of 1,1-dichloromethylenes are formed. Evidently chlorination of the structural monomer units of the polymer reduces considerably the vinylidene chloride copolymer sequences when compared with those of the 1,2-dichloroethylene copolymer. Chlorine preferentially attacks the central carbon atom of heterotactic triads [3]. Increasing the number of chlorine atoms in a structural unit [4] gradually decreases the rate of chlorination. Chlorination is a radical chain reaction proceeding via repeating cycles of the transfer of a chlorine atom to a C—H bond

$$\sim CH_2CHCl \sim + \; Cl^{\cdot} \;\; \rightarrow \;\;\; \sim \dot{C}HCHCl \sim + HCl$$

and of the subsequent reaction of the resulting alkyl radical with molecular chlorine

$$\sim \dot{C}HCHCl \sim + \; Cl_2 \;\; \rightarrow \;\;\; \sim CHClCHCl \sim + Cl\cdot$$

The regenerated chlorine atom starts a new propagation step and the chain reaction proceeds until all radical centres disappear. Termination occurs via the cross reaction of chlorine atoms and alkyl radicals. Since an insoluble crosslinked polymer gel is formed during the chlorination of polyethylene, the self-recombination of alkyl radicals is also of importance. Bromination is somewhat different since it yields soluble brominated polyethylene. The relatively faster reaction of alkyl radicals with molecular bromine (than with chlorine) reduces the stationary concentration of alkyl radicals in the system and the probability of their dimerization is consequently lower.

Chlorinated polyethylene finds practical application as an additive which increases the toughness of poly(vinyl chloride). Chlorinated poly(vinyl chloride) has a higher softening temperature. (Commercial chlorinated poly(vinyl chloride) contains up to 66 % by weight chlorine.)

Chlorination also increases the thermal stability of poly(vinyl chloride). This may be explained by the addition reaction of chlorine which eliminates some double bonds, which are the potential sites for the initiation of dehydrochlorination [5]. Since the poly(vinyl chloride) macromolecule may involve several types of anomalous structural units, chlorination itself is not usually sufficient to obtain a perfectly thermally stable polymer, but in combination with other means of polymer stabilization it may be of use.

When polyethylene is treated with chlorine and sulphur dioxide, chlorosulphonyl groups may be obtained on the polymer chain

$$\dot{R} + SO_2 \;\; \longrightarrow \;\; R\dot{S}O_2 \;\; \xrightarrow{Cl_2} \;\; RSO_2Cl + Cl\cdot$$

Chlorosulphonation of polyethylene as conducted commercially gives a product with about 27 % (by weight) chlorine and 1.5 % (by weight) sulphur. Incorporation of chlorosulphonyl groups into the polymer reduces the crystallinity and the modified polymer has rubber-like properties; ZnO is usually used as a vulcanization coagent, acting by forming crosslinks with pendant groups. The chlorosulphonyl groups also suppress the flammability of the polymer. Their subsequent hydrolysis to sulphonic acid groups finds use in the preparation of cation-exchange resins.

The reduction of their paraffinic character by the insertion of polar groups into hydrocarbon polymers may be effected by phosphorus chlorides or oxychlorides

$$\sim CH_2 \sim + 2\,PCl_3 \xrightarrow{\;O_2\;} \sim CH \sim + HCl + POCl_3$$
$$\underset{\displaystyle O = PCl_2}{\big|}$$

or by oxidants such as nitric or bromic acids or their mixtures. The nonselective action of these acids leads to the appearance of alcoholic, ketonic, aldehyde and carboxyl groups on a hydrocarbon chain as well as to rapid degradation of the chain. Because of the complex mechanism, oxidation by these acids has not been investigated theoretically but it is used in practice in the surface modification of polymers, making them more hydrophilic.

a. Reactions of Double Bonds in Hydrocarbon Polymers

Successive reactions of double bonds in a polymer chain occur very easily [6]. Of the various possibilities, *cis-trans* isomerization of a carbon-carbon double bond on a macromolecular backbone is structurally the simplest.

The mechanism of the *cis-trans*-isomerization of 1,4-polybutadiene photosensitized by organic sulphides and bromides involves the formation of a transient free radical

By the addition of radical X, the double bond of the *cis*-isomer is converted to a single bond. Free rotation about this bond and subsequent elimination of radical X leads to the formation of the *trans*-isomer. Isomerization occurs via a chain reaction of the considerable length. Each thiophenyl radical $C_6H_5S^{\cdot}$ is capable of initiating the isomerization of about 1200 *cis*-double bonds before its disappearance in some termination reaction. The kinetic chain length for bromine atoms is 350.

The isomerization of 1,4-polybutadiene depends on the initial and equilibrium amounts of the isomers. The latter depends on the structure of the polydiene and the experimental conditions (the mode of initiation, pressure, temperature, etc.). At equilibrium, 1,4-polybutadiene comprises 33 and 67 mol % of the *cis*- and *trans*-isomers respectively. The photosenzitized reaction of the *cis*-isomer gives, therefore, polymer enriched with *trans*-isomer units. On the other hand, irradiation of *trans*- 1,4-polybutadiene gives rise to *cis*-isomer units. Similar equilibrium values for the individual isomers may also be achieved by prolonged heating (e. g. 230 °C, 100 h for of *cis*-1,4-polybutadiene in vacuo). Gamma-irradiation produces a level of the *cis*-isomer of 20 mol %.

Isomerization of *cis*-1,4-polybutadiene also occurs during its vulcanization by sulphur at 140—160 °C and in other radical reactions. If the hydrogen on the carbon-carbon double bond of polybutadiene is replaced by a methyl group, as in 1,4-polyisoprene, then isomerization fails to occur under the same experimental conditions. Isomerization of natural rubber (which is in fact 1,4-polyisoprene) is still possible at 180—200 °C in the presence of selenium when the resulting *cis-trans* equilibrium is about 1 : 1.

Cyclization provides another route for the isomerization of polydienes which reduces the unsaturation of the polymer while keeping the C : H ratio constant. Strong acids (H_2SO_4, HCl, some Lewis acids, etc.) convert natural rubber or other polydienes to hard, nonrubber-like, thermoplastic materials of higher density than the parent polymer. This transformation may be attributed to an intramolecular cationic polymerization

The process proceeds by a repetition of the above steps and via addition of other unsaturated mers to form cyclohexene structural units, unless prevented by steric hindrance

Since the cycloaddition occurs randomly at structural units of the polymer chain, some mers do not participate in the reaction and remain isolated (*Fig. 3.1*).

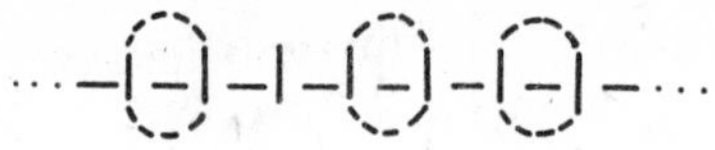

Fig. 3.1. Formation of an isolated isoprene (I) mer in polymeric segments of cyclized natural rubber

Cyclization of 1,2-polybutadiene [6] gives cycloalkane structural units of various structures

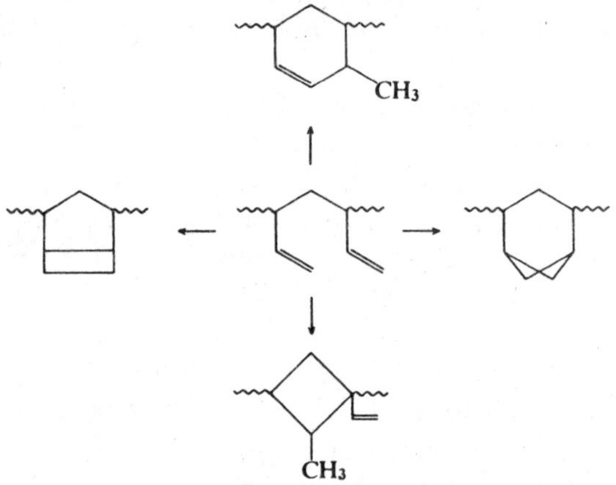

The reaction, which proceeds by a radical mechanism, is initiated by heating the polymer to 100—300 °C in an inert atmosphere. Its activation energy is 142 kJ mol^{-1}. Similar cycloalkane structural units are also formed from anomalous mers of 1,2-polybutadiene; only the position of the rings attached to the main chain is different. The reduction of unsaturation in 1,4-polybutadiene due to cyclization when compared to the 1,2-isomer, is considerably slower. A similar type of cyclization reaction also occurs with other 1,2- or 3,4-polydienes.

In the presence of organic peracids, the *epoxidation* of a double bond of a polydiene [7]

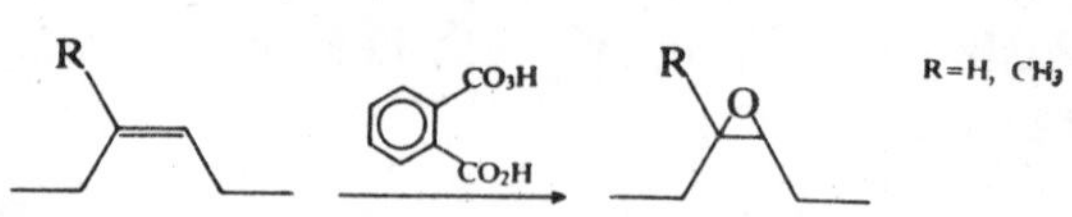

proceeds in a similar way to low-molecular-mass alkenes. The reactivity of the double bond in polydiene structural units towards the peracid increases both with the degree of substitution and with the accessibility of the double bond as follows

with a *cis*-C=C bond being more reactive than its *trans*-isomer.

60

The formation of reactive oxirane rings on the macromolecular backbone as a result of epoxidation facilitates subsequent *functionalization* of a polymer. Reaction of polymeric oxiranes with long-chain dicarboxylic acids gives elastic thermosets; a shorter separation of the carboxyl groups in di-acid leads to the typically harder products of a crosslinked polydiene. Hydroxyl and carboxyl groups present in the polymer as side products of epoxidation may, however, effect crosslinking during the epoxidation itself, although this usually takes place at high conversions.

Although epoxidized natural rubber is currently available only in small tonnage quantities as a development product, these materials have considerable potential as epoxy polydiene resins [8].

Polydienes may be also modified via addition reactions to other polar compounds. Many papers are devoted to the addition of maleic anhydride to natural rubber, which may be initiated thermally

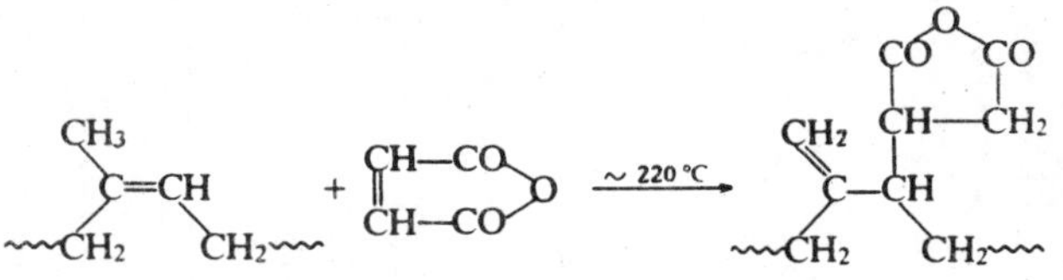

by some free radical initiators or by light. In addition to the mode of attachment of maleic anhydride to the 1,4-*cis* polyisoprene unit shown above, there are other possibilities (a)—(c)

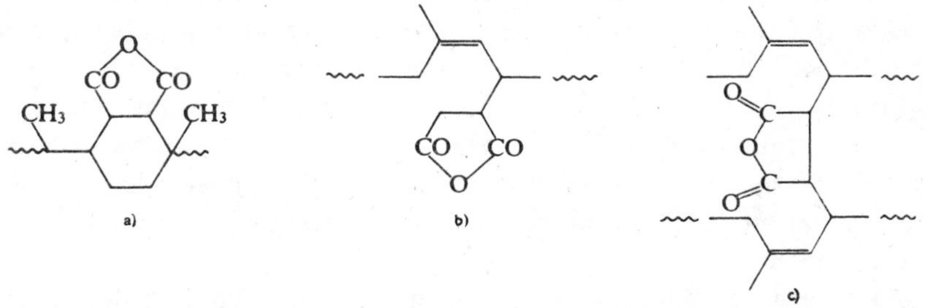

Product (c) explains the crosslinking of rubber at higher fractions of bound maleic anhydride. The subsequent crosslinking of rubber by difunctional amines and metal oxides requires about 5% of anhydride. Maleic anhydride reacts with polydienes very easily; the analogous reaction can be effected with the aromatic rings of polystyrene on photochemical initiation [9].

As illustrated by the following examples, the incorporation of ionic groups into unsaturated hydrocarbon polymers may be achieved by a radical addition of a suitable reagent followed by subsequent hydrolysis and neutralization of the product [10].

$$-CH=CH- \quad \xrightarrow[\text{2. } H_2SO_4 \quad \text{3. NaOH}]{\text{1. } HSCH_2COOCH_3} \quad -CH_2CH-$$
$$\quad\quad\quad\quad\quad\quad\quad\quad\quad\quad\quad SCH_2COONa$$

$$\xrightarrow[\text{2. HCl } \quad \text{3. NaOH}]{\text{1. } (CH_3O)_2POH} \quad -CH_2CH-$$
$$\quad\quad\quad\quad\quad\quad\quad\quad PO(ONa)_2$$

$$\xrightarrow[\text{2. NaOH}]{\text{1. } SO_3 \text{ in}(C_2H_5O)_3PO} \quad -CH_2CH-$$
$$\quad\quad\quad\quad\quad\quad\quad SO_3Na$$

Such reactions lead to a physical crosslinking via aggregates of ionized groups in the matrix of the hydrocarbon polymer, and considerable changes in properties are achieved at net conversions of only 5 mol %.

Polydienes having reactive allylic hydrogens also react with other electrophiles. For example, the addition reaction of 4-phenyl-1,2,4-triazoline-3,5-dione and 1,4-polyisoprene [11]

is complete in a few minutes at 20 °C.

The presence of a polar group in a hydrocarbon chain increases the glass transition temperature of the polymer as well as its impact strength.

The halogenation of polydienes and especially of natural rubber was carried out in connection with attempts to explain the chemical structure and macromolecular character of these polymers. Despite much experimental data, this field is still primitive from the theoretical viewpoint. This is because the halogenation of polydienes is a complex process where reactive intermediates also initiate cyclization and crosslinking. The halogenation of rubbers improves their resistance to ozone.

From the spectrum of other modification reactions of polydienes, the hydrochlorination of natural rubber, which finds practical application is noteworthy.

Hydrogenation has not been exploited industrially to date but it may prove useful in the modification of the properties of triblock copolymers of polydienes. The hydrogenation catalyst is platinum black [12]. Hydrogenation of polybutadiene in heptane solution occurs under one atmosphere of hydrogen at 20 °C. Completely hydrogenated polybutadiene has properties identical with linear low-pressure polyethylene. Partially hydrogenated polybutadiene, corresponding to a copolymer of butadiene and ethylene has, however, a different composition from material prepared by direct copolymerization. Different levels of unsaturation in a copolymer bring about different crystallinity, elasticity and also chemical reactivity, particularly as regards resistance to weathering.

The radical addition of aromatic thiols is used especially for the plasticizing of rubbers. Plasticizing results from macromolecular scission accompanied by the formation of free radicals which, after reaction with molecular oxygen and the formation of peroxyl radicals, abstract hydrogen from the thiol group. The resulting arylthiyl radical then adds to a double bond in rubber. The alkyl radicals thus produced react once more with oxygen to give peroxyl radicals which undergo fragmentation to thiyl radicals and peroxide

$$R^1CH{=}CHR^2 + R^3S^{\cdot} \rightarrow (R^1)(R^3S)CH{-}\dot{C}HR^2$$

$$\underset{(R^1)(R^3S)CH\dot{C}HR^2}{\overset{\overset{\textstyle OO^{\cdot}}{|}}{}} \longrightarrow \underset{R^1CHCHR^2}{\overset{\overset{\textstyle O-O}{|\quad|}}{}} + R^3S^{\cdot}$$

By addition to a double bond of polydiene, the thiyl radicals enter a new reaction cycle. The reaction intermediate featuring a 1,2-dioxetane unit is unstable and decomposes, a process being associated with the reduction in the molecular mass of rubber.

When compared with aromatic thiols, the addition of aliphatic thiols to polydienes occurs to a much greater extent in the presence of air, but without significant scission of the polymer. The highest rate of addition was observed for the *cis*-isomers. Addition of aliphatic thiols improves the resistance of rubbers towards ozone and reduces the solubility and permeability of gases.

In the first stage of *addition of ozone* to the double bonds [13] of polydienes, ozonides and cyclic peroxides are formed

In water, these unstable addition products are quickly decomposed to ketones and aldehydes

and the polymer chain is disrupted, to give a decrease in its average molecular mass. When compared with the single C—H and C—C bonds of saturated chains (*Table 3.1*) [14], the attack of ozone on a double C=C bond is about six orders of magnitude faster. The rate constants of this reaction are approximately the same for solid polydiene and for its tetrachloromethane solutions [15]. The unusually fast reaction of polydienes with ozone even at ambient tem-

Table 3.1   Rate Constants and Relative Rates (v_{rel}) for the Reaction of Ozone with Hydrocarbon Polymers

Polymer	Structural mer	$k^{a)}$/ dm^3 mol^{-1} s^{-1}	v_{rel}
Polyisobutylene	$|-CH_2C(CH_3)_2|n$	1.2×10^{-2}	1
Polyethylene	$|-CH_2-|n$	4.6×10^{-2}	3.8
Polyphenylene	$|-C_6H_4-|n$	5.0×10^{-2}	4.2
Polystyrene	$|-CH_2CHC_6H_5|n$	0.3	25
Polyvinylcyclohexane	$|-CH_2CHC_6H_{11}|n$	0.8	67
Polycarbonate	$|-C_6H_4C(CH_3)_2C_6H_4OCO_2-|n$	3.0	250
Polyphenylacetylene	$|-CH=CC_6H_5|n$	$1.4 \times 10_3$	1.2×10^5
Polybutadiene	$|-CH_2CH=CHCH_2-|n$	6.0×10^4	5.0×10^6
Polyisoprene	$|-CH_2CH=C(CH_3)CH_2-|n$	1.4×10^5	1.2×10^7

a) Rate constant for 20 °C in CCl₄

peratures underlines the necessity of protecting these polymers against ozone. A high reactivity of a double bond towards ozone is of course also characteristic for low-molecular-mass compounds, but in macromolecular compounds, however a special kind of "polymer effect" is apparent featuring a dependence of the reaction rate on the degree of mechanical deformation of the polymer. Macromolecular radicals generated by the fracture of polymer molecules under the action of frictional shearing forces undergo reaction even with metals [16]. This reaction can account for the wear of hard metals by soft elastic solids.

Among the reactions of conjugated double bonds of a macromolecular backbone, of special note are the *processes of doping by strong electron donors* (naphthalenides of the alkali metals in tetrahydrofuran, eutectic mixtures of K and Na, etc.) and *electron acceptors* [17] (AsF_5, I_2, Br_2, $SbCl_5$, HSO_3F, SO_3, IF_5, etc.). Such doping increases the electrical conductivity of a polymer by up to 16 orders of magnitude, i.e. to approach metallic conductivity.

The simplest example of a polymer with conjugated double bonds is polyacetylene; in its interaction with alkali metals M (Li, Na, K, Rb or Cs), a complex of maximum possible uptake expressed by the formula $(C_3H_3M)_n$ is formed. In structures with M = Li about 30 % of an electronic charge is localized on the carbon atoms [18]

Polyacetylene doped with halogens has an even higher electrical conductivity. Delocalized cations are formed on the polymer chain which are compensated by the negative charge of I_5^-, I_3^-, Br_3^-, or Cl_3^- ions. Modified polyacetylene undergoes simultaneous halogenation by substitution at its side-groups, with relative loss of conductivity. Doping with electron donors and acceptors alters the conformation of polyacetylene from *cis* to *trans*. For polyacetylene, the highest conductivity (1200 S cm^{-1}) was achieved experimentally with AsF_5 as dopant.

A similar, high conductivity of a polymer complex may be obtained with poly(*p*-phenylene). The progress of doping of poly(*p*-phenylene) can be monitored via the various colour changes: while the original poly(*p*-phenylene) is brown, the incorporation of gaseous AsF_5 turns the polymer green-black whereas interaction with Na—K alloy yields a golden material. The ratio of hydrogen to carbon in the polymer remains constant. The degree of undesired side-reactions depends very much on the reaction conditions and on the quality of the reactants [19]. The electrical conductivity of doped poly(*p*-phenylene) temporarily disappears on gradual replacement of the electron donor by an electron acceptor or vice versa until above the stoichiometric 1 : 1 ratio of donor and acceptor, the mutual deactivation of the dopants ceases and the conductivity increases once again.

All doped polyconjugated polymers are very moisture-sensitive through elimination of the reactive charge-transfer complexes, which reduces the conductivity to values typical for the original polymer. All manipulations with doped polymers should therefore be carried out in a strictly inert medium.

The process of doping is usually reversible, a phenomenon used in practice in chargeable electric cells or batteries [20]. Undoped poly(*p*-phenylene) is a good electrical insulator having a conductivity lower than 10^{-12} S cm^{-1}. At the beginning of its electrochemical doping, the conductivity of the polymer is slightly increased by AsF_5. Provided that tetraethylammonium hexafluorophosphate in perfectly dry propylene carbonate is used as an electrolyte, the polymer electrode should be connected to the positive pole of d.c. voltage source. Poly(*p*-phenylene) is then gradually doped with PF_6^- anions and its conductivity increases up to 50 S cm^{-1}. Electron-donor doping with lithium is performed with a lithium electrode immersed in a solution of sodium perchlorate in tetrahydrofuran.

The doping process corresponds to the discharge (1) of the battery

$$(—C_6H_4—)_n + anLi \underset{2}{\overset{1}{\rightleftharpoons}} [(—C_6H_4—)^{-a}Li^{+a}]_n$$

while the reversible reaction corresponds to its charging (2). In this case, the stoichiometric coefficient a for discharging is 0.15. The battery incorporating lithium hexafluoroarsenate offers some advantages. During its charging (3)

$$(-C_6H_4-)_n + anLiAsF_6 \xrightleftharpoons[4]{3} [(-C_6H_4-)^{+a} (AsF_6)^{-a}]_n + anLi$$

poly(p-phenylene) is oxidized and AsF_6^- anions of the electrolyte enter the polymer electrode where they function as counter-ions to the polymer cations. During discharge (4) these anions diffuse back into the electrolyte.

Such doping reactions may also be performed with other polyconjugated polymers, and the application of these processes to new types of batteries offers only one of many potential growth areas in this particular area of macromolecular chemistry [21, 22, 23].

b. Reactions of Phenyl Substituents on a Polymer Chain

The functionalization of polystyrene is a typical example of the applications of organic chemistry to macromolecular systems. The *sulphonation* of crosslinked copolymers of styrene and divinyl benzene by concentrated sulphuric acid is, for example, used in the preparation of cation-exchange resins and specific reversible adsorbents or polymer catalysts [24].

The great excess of H_2SO_4 in the reaction medium ensures the binding of one sulphonate group to each structural unit of the copolymer. From noncrosslinked, slightly sulphonated polystyrene with 5 mol % of sulphonate groups, the physically crosslinked polymer may be obtained after neutralization of sulphonic acid groups with metal cations. Such a polymer can be moulded even at 220 °C.

Attachment of a *nitro group* to the aromatic ring of polystyrene is also quite easy. The nitro group can be successively reduced to an amino group by phenylhydrazine. The basic *amino group* can then serve as a scavenger of acids or of certain metal ions, and acts as a precursor for the preparation of new polymer reagents.

Chloromethylation [25, 26] by chloromethyl methyl ether (as a mixture of CH_2O and HCl) catalyzed by $SnCl_4$, $ZnCl_2$ or boron trifluoride etherate provides a successful method of functionalization of crosslinked polystyrene

Chloromethylations are accompanied by a side reaction in which methylene bridges link phenyl groups; such crosslinking may be promoted by higher concentrations of $AlCl_3$. The chloromethyl groups may be transformed to aldehydes and, after subsequent oxidation, to carboxyl groups

$$\text{(P)}\!\!-\!\!\langle\text{C}_6\text{H}_4\rangle\!\!-\!\!\text{CH}_2\text{Cl} \xrightarrow[\text{NaHCO}_3]{\text{(CH}_3)_2\text{SO}} \text{(P)}\!\!-\!\!\langle\text{C}_6\text{H}_4\rangle\!\!-\!\!\text{CHO} \xrightarrow{\text{O}_2} \text{(P)}\!\!-\!\!\langle\text{C}_6\text{H}_4\rangle\!\!-\!\!\text{COOH}$$

Nucleophilic substitution of a chloromethylated derivative provides a variety of possibilities for the functionalization of a polymer

$$
\begin{array}{ll}
\xrightarrow{(CH_3)_2S} & \text{(P)}-CH_2\overset{\oplus}{S}(CH_3)_2\overset{\ominus}{Cl} \\
\xrightarrow{KSCH_3} & \text{(P)}-CH_2SCH_3 \\
\xrightarrow{RCOONa} & \text{(P)}-CH_2OCOR \\
\xrightarrow{KCN} & \text{(P)}-CH_2CN \\
\xrightarrow{NR_3} & \text{(P)}-CH_2\overset{\oplus}{N}R_3\overset{\ominus}{Cl} \\
\xrightarrow{SC(NH_2)_2} & \text{(P)}-CH_2SC{=}NH(NH_2),\ \text{(P)}-CH_2SH \\
\xrightarrow{NH_3} & \text{(P)}-CH_2NH_2 \\
\xrightarrow{LiPPh_2} & \text{(P)}-CH_2PPh_2 \\
\xrightarrow{AlBr_3} & \text{(P)}-CH_2\overset{\oplus}{AlBr}\overset{\ominus}{Cl} \\
\xrightarrow{P(OEt)_3} & \text{(P)}-CH_2P(O)\,(OEt)_2
\end{array}
$$

The substitution of chlorine in the chloromethyl group is, of course, open to reagents other than indicated in the scheme. The nonequivalence of individual chloromethyl groups in crosslinked polystyrene should be noted. If we allow an aminoacid to react with crosslinked poly(chloromethylstyrene)

$$\text{(P)}-CH_2Cl + H_2NRCO_2H \longrightarrow \text{(P)}-CH_2NHRCO_2H$$

then fewer functional groups react than when the polymer has been pre-treated with dimethyl sulphide. This may be explained by the binding of dimethyl sulphide to poly(chloromethyl styrene) through formation of the more polar dimethyl-sulphonium salts

$$\text{(P)}-CH_2Cl + (CH_3)_2S \longrightarrow \text{(P)}-CH_2\overset{\oplus}{S}(CH_3)_2\overset{\ominus}{Cl}$$

which facilitate the diffusion of the aminoacid through the polymer matrix. The aminoacid may thus penetrate to more deeply located reaction sites and a greater degree of reaction is attained than when reaction is largely confined to the surface of the particles of crosslinked polymer. The apparent reactivity of the functional groups also depends on the size of the polymer particles, their porosity, and on the dimensions of the pores as well as on reacting molecules themselves and other factors [27].

Even though chloromethylated polystyrene is undoubtedly a key polymer reagent, there are other examples of the functionalization of polystyrene which have considerable research potential [28]. Many potential applications emanate from *the interaction of lithium with the phenyl groups* followed by deactivation of organolithium polymer with a suitable electrophile

The reaction with oxygen transforms the lithiated phenyl groups of polystyrene to hydroxylated polystyrene which can act as the starting compound of numerous polymer-analogous reactions [29]. CO_2 substitutes lithium by carboxyl groups, while reaction with Ph_2PCl gives the polymeric analogue of triphenyl phosphine.

Polystyrene may also be modified through isomerization of stereoregular polystyrene chains. The action of *tert-* butylpotassium in dimethyl sulphoxide for several tens of hours converts the isotactic polymer to its atactic isomer. Such a change in the configuration of substituents may be accomplished only by strong basic reagents capable of abstracting a proton from the polymer

Thus it is the polymer anion which undergoes isomerization [30].

68

The same kind of spatial rearrangement was observed for poly(alkyl methacrylates) and it may even occur with saturated hydrocarbon polymers. Here the problem of achieving simultaneously reasonable solubility of the nonpolar polymer and polar reagent in a single solvent complicates the matter considerably.

Since the atactic isomers which are the only products of such transformations are readily available via direct synthesis, the above epimerization reactions are mainly of theoretical interest.

B. REACTIONS OF THE HETEROATOMS OF SIDE GROUPS

The presence of oxygen attached to a hydrocarbon chain in the form of an ester, ether, carbonyl or hydroxyl group increases the reactivity of the macromolecule considerably. From the viewpoint of practical applications, the synthesis of poly(vinyl alcohol) should be mentioned first. It should be recalled that poly(vinyl alcohol) can be prepared only by a substitution reaction conducted on the mers of a macromolecular chain. (Monomeric vinyl alcohol is unstable and quickly rearranges to acetaldehyde.) Poly(vinyl alcohol) is usually prepared by *the hydrolysis* of poly(vinyl acetate)

$$\left[\begin{array}{c} -CH_2CH- \\ | \\ OCOCH_3 \end{array}\right]_n \longrightarrow \left[\begin{array}{c} -CH_2CH- \\ | \\ OH \end{array}\right]_n$$

Poly(vinyl alcohol) finds numerous technical applications. Since its hydroxyl groups can be easily transformed to ether or ester groups it may also be used as a reagent for the preparation of other polymers. The solubility of poly(vinyl alcohol) in water as well as its other properties depends considerably on the residual content of acetate groups [31].

The low temperature esterification of the hydroxyl groups of poly(vinyl alcohol) with cinnamoyl chloride

$$-(CH_2CH)_n + n\ C_6H_5CH=CHCOCl \longrightarrow -(CH_2CH)_n$$
$$| |$$
$$OH OCOCH=CHC_6H_5$$

carried out in the presence of pyridine, which removes the liberated HCl, is of particular importance. The reaction is a fundamental process in the preparation of photocrosslinked polymers (resists) widely used in microelectronics. The best photolithographic properties are achieved when the degree of esterification reaches about 95 % and for polymers having as narrow a polydispersity as possible. Increasing the molecular mass enhances the sensitivity of a polymer to ultraviolet light since fewer photons per volume unit are necessary to induce crosslinking of the photoexposed polymer layer. The crosslinks between the macromolecules are formed by cyclodimerization of the double bonds of the cinnamic acid moieties attached to different macromolecules. The cyclobutane ring crosslinkages arise only from the *trans*-isomers [32, 33].

The acid-catalyzed condensation of poly(vinyl alcohol) and aldehydes such as formaldehyde, acetaldehyde and butyraldehyde is important in the production of poly(vinyl acetals) which are speciality products with a low tonnage production. Assuming random isolation of unreacted hydroxyl groups, the maximum degree of acetalation should not exceed 86.5 mol %. The yields found experimentally are, however, higher. Since any reversibility of acetalation has not been confirmed experimentally, it appears to proceed as a non-random zipper-like process where the condensation at neighbouring hydroxyls is controlled by pre-existing acetal groups [34].

The substitution of the hydrogens of the hydroxyl groups of poly(vinyl alcohol) by the sodium atom of CH_3SOCH_2Na to yield alkoxides

$$\left[-CH_2CH-\right]_n \xrightarrow{CH_3SOCH_2Na} \left[-CH_2CH-\right]_n$$
$$\quad\quad\; OH \quad\quad\quad\quad\quad\quad\quad\quad ONa$$

provides a further possibility for modification of the properties of poly(vinyl alcohol). Addition of propane sulphone to alkoxide groups gives alkyl sulphonate groups attached to the main chain via oxygen atoms

$$\left[-CH_2CH-\right]_n \xrightarrow{\overset{(CH_2)_3-SO_2}{\underset{O}{\vert}}} \left[-CH_2CH-\right]_n$$
$$\quad\quad\; ONa \quad\quad\quad\quad\quad\quad\quad\quad O(CH_2)_3SO_3Na$$

Alkyl halides featuring long alkyl groups may thus be introduced into poly(vinyl alcohol) to provide crystallizable substituents [35].

The reaction of two neighbouring hydroxyl groups of the poly(vinyl alcohol) chain with cyanogen bromide is also of importance since

$$\left[\begin{matrix}-OH\\[4pt]-OH\end{matrix}\right. \xrightarrow{BrCN} \left.\begin{matrix}-O\\[4pt]-O\end{matrix}\right\rangle C=NH \xrightarrow{H_2N-X} \left[\begin{matrix}-OCONHX\\[4pt]-OH\end{matrix}\right.$$

it may be used for the binding of proteins through their amino groups to polyhydroxylated polymers.

The most convenient means of *attaching nucleophilic reagents* to the hydroxyls of poly(vinyl alcohol) or cellulose is provided by the re-esterification exchange reaction. On the condensation of tosyl chloride with the hydroxyl group, an unstable ester is primarily formed

70

$$P\text{—OH} + ClSO_2 \langle \rangle CH_3 \longrightarrow P\ OSO_2 \langle \rangle CH_3 + HCl$$

from which the tosyl groups may be replaced by a nucleophilic reagent, $Nu^-(RS^-, CH_3COO^-, C_6H_5O^-, \text{etc.})$

$$P\ OSO_2 \langle \rangle CH_3 + Nu^- \longrightarrow P\text{—Nu} + {}^-OSO_2 \langle \rangle CH_3$$

Re-esterification does not always proceed as smoothly as in the above case but it may also be conducted with esters of polyacids and low-molecular-mass alcohols. Hydrolysis of the ester group in polymethacrylates is catalyzed by adjacent carboxyl anions; the kinetics of the reaction show distinct autocatalytic character. The adjacent mers do not accelerate reaction but may however slow it down. The rate of the alkaline hydrolysis of polyacrylamide is, for example, reduced more markedly than expected from the decrease of the concentration of reacting functional groups. As it was pointed out earlier, this may be due to electrostatic interactions of neighbouring groups.

In the reactions of organolithium compounds with normal ester groups, the substitution of alkoxyl at the carbonyl group occurs to yield ultimately a tertiary alkoxide group.

$$[-CH_2\overset{|}{C}CH_3]_n \xrightarrow{RLi} [-CH_2\overset{|}{C}CH_3]_n \qquad (R = PhSCH_2; PhCH_2)$$
$$CH_3O\overset{|}{C}O \qquad\qquad R\overset{|}{C}O$$

$$[-CH_2\overset{|}{C}CH_3]_n \xrightarrow{RLi} [-CH_2\overset{|}{C}CH_3]_n \qquad (R = CH_3(CH_2)_3$$
$$CH_3O\overset{|}{C}O \qquad\qquad R_2\overset{|}{C}O^-Li^+$$

Apart from these two reactions, the ester group of poly(methyl methacrylate) may also undergo cyclization

$$[-CH_2\overset{|}{C}CH_3]_n \xrightarrow{PhLi} [-CH_2\overset{|}{C}CH_3]_n \xrightarrow{PhLi} [-CH_2\overset{|}{C}(CH_3)CH_2\overset{|}{C}CH_3]_m \longrightarrow$$
$$CH_3O\overset{|}{C}O \qquad\qquad Ph\overset{|}{C}O \qquad\qquad (Ph)_2\overset{|}{C}\underset{O}{\diagup\diagdown}\overset{|}{C}{=}O$$

$$\xrightarrow{PhLi} [-CH_2\overset{|}{C}(CH_3)CH_2\overset{|}{C}CH_3]_m$$
$$(Ph)_2\overset{|}{C}\underset{O}{\diagup\diagdown}Ph\overset{|}{C}\text{—}O^-Li^+$$

which happens particularly with an excess of phenyl lithium.

From the practical viewpoint, the Hoffman elimination of the carbonyl group of polyacrylamide [36] is of interest.

71

$$[-\underset{\underset{CONH_2}{|}}{C}HCH_2-]_n \xrightarrow[\text{2. HCl}]{\text{1. Cl}_2\text{(NaOH)}0°C} [-\underset{\underset{NH_2.HCl}{|}}{C}HCH_2-]_n$$

Reaction proceeds with a high yield of poly(vinyl amine) which cannot be prepared in any other way. As in the case of vinyl alcohol, vinyl amine is unstable and cannot be polymerized directly. Poly(vinyl amine) is readily soluble in water. Its reactive amino groups can be used for the binding of biologically active compounds. As regards the reactions of the amino groups, their neutralization with acids is noteworthy. The primary chemical structure of the mers induces such conformations of the macromolecular chain which correspond to the sizes of the molecules of the surrounding medium. Adducts of some acids with the amino groups are so unstable that the acid groups (when there is an excess of amino groups) may migrate along the chain. The process may be initiated merely by altering the solvent which changes the secondary structure of the polymer chain.

The carbonyl group present as an inherent part of the main chain in aliphatic polyketones (copolymers of CO and ethylene) may be transformed by ammonia to an amide link

$$-CH_2CH_2\overset{\overset{O}{\|}}{C}- \xrightarrow{NH_3} -CH_2CH_2CONH-$$

and the hydrocarbon backbone of the parent polymer is thereby changed to that of a heteroatom-based macromolecule.

The presence of a halogen attached to the structural units of polymers enables reactions with basic amines and with phosphines

$$[-CH_2\overset{|}{C}HCl]_n + n\ (Ph)_3P \longrightarrow [-CH_2\overset{|}{C}H]_n$$
$$(Ph)_3\overset{\oplus}{P}Cl^{\ominus}$$

The reaction is accompanied by dehydrohalogenation which is the most frequent process causing deterioration in the properties of chlorinated and brominated polymers.

Despite numerous attempts, the complete *dehydrohalogenation* of poly(vinyl chloride), a target oriented towards the preparation of electroconductive poly-conjugated polymers, has not been achieved. This is probably associated with the fact that the dehydrochlorination reaction is insufficiently selective which, together with the presence of anomalous structural units, hinders the formation of the required product. Polyacetylene gives a higher electrical conductivity and more reproducible results and its use is therefore preferred.

Another way of modifying poly(vinyl chloride) involves its *dechlorination*. When zinc powder interacts with poly(vinyl chloride) solution, the final product of the incomplete process involves cyclopropyl (49%) and, to a lesser extent, cyclopentyl rings, and a relatively high concentration (21 mol %) of methylene mers and double bonds (9 mol %) [37]. The unexpected appearance of methylene units in place of CHCl groups is probably due to the transfer reaction of alkyl radicals to hydrogen-containing molecules of the solvent. The complete dechlorination of the polymer occurs with tributyltin hydride in a chain reaction

$$(C_4H_9)_3Sn^{\cdot} + ClCHR^1R^2 \;\rightarrow\; (C_4H_9)_3SnCl + \dot{C}HR^1R^2$$

$$\dot{C}HR^1R^2 + HSn(C_4H_9)_3 \;\rightarrow\; CH_2R^1R^2 + \dot{S}n(C_4H_9)_3$$

This radical reaction may be initiated by azobis(isobutyronitrile), the primary cyanopropyl radicals of which can abstract hydrogen from tributyltin hydride. They are not, however, capable of transfer reactions to parent or dechlorinated poly(vinyl chloride).

Allylic chlorine atoms in the polymer chain may be removed by an exchange dehalogenation reaction

This may explain the effect of some thermal stabilizers of poly(vinyl chloride) (such as salts of tin, lead, cadmium, calcium and zinc with organic acids) as well as the synergistic effect of epoxy compounds

In addition to these processes, allylic chlorine is reactive towards aqueous alcohol [38] which substitutes chlorine by the more stable hydroxy or alkoxy group. Exchange reactions are also used to substitute chlorine bound to tertiary carbon atoms, which are the sites of branching in poly(vinyl chloride). Improvement in the least desirable property of poly(vinyl chloride), namely its relatively low thermal stability, by a simple and efficient elimination of labile chlorine atoms, immediately following polymerization, would find widespread utilization.

73

C. REACTIONS OF POLYMERS CONTAINING HETEROATOMS

The presence of heteroatoms in the backbone of a macromolecule increases the number of possible reactions of the relevant structural units. Such factors such as the relative reactivity of the side groups and the heteroatoms in the main chain, the character of the reagents used, etc., determine which of the possible reactions will predominate. New pathways for the transformation of heterochain polymers could involve interactions of the side groups with the heteroatoms of the main chain. One such reaction is *the isomerization of the structural units of polymers* having sulphur in the chain and a pendant halogenoalkyl group [39]

$$-CH_2CHS- \ \rightleftharpoons \ -CH_2CHCH_2S-$$
$$\quad\quad | \quad\quad\quad\quad\quad\quad\quad |$$
$$\quad CH_2Cl \quad\quad\quad\quad\quad\quad Cl$$

After two months of storage in solution at 35 °C, each of the above structures gives approximately the same equilibrium concentration of both structural mers of the resulting copolymer.

Photoisomerization of some aromatic polycarbonates to yield hydroxybenzophenone units is another example of this kind of reaction

Its course is autoinhibited by the hydroxybenzophenone structural units produced which absorb the incident ultraviolet light and transform it to thermal energy. Photoisomerization to produce photostabilizing moieties is in accordance with the very good ultraviolet light-resistance of some polycarbonates [40].

The transformation of polyamides to polyesters is an example of a two-step reaction which starts at the pendant group and continues at the heteroatoms of the main chain. The first step is the nitrosation of the amido group. The product formed is thermally decomposed to yield an ester group as part of the polymer chain

$$P^1CONHP^2 \ \xrightarrow[-H_2O]{N_2O_3} \ P^1CONP^2 \ \xrightarrow{-N_2} \ P^1COOP^2$$
$$\quad\quad\quad\quad\quad\quad\quad\quad\quad |$$
$$\quad\quad\quad\quad\quad\quad\quad\quad\quad NO$$

Reaction of a suitable side group with the heteroatoms of the main chain of aromatic polyamides provides a route to polyimides

Similarly polyoxadiazoles can be produced from polyhydrazides

In the presence of amines the latter reaction yields polytriazoles

Thermostable aromatic polyamides, polyoxadiazoles and some polytriazoles are insoluble in any solvent and do not melt on heating. Any modification reaction of an appropriate polymer precursor should, therefore, be carried out on material in the shape of form desired in the final product, such as a foil or fibre.

The oxidation of poly(hexamethylene sulphide) by hydrogen peroxide [41] is an example of a nondestructive reaction taking place on the heteroatoms of the main chain. The oxidation is a two-step process. At lower concentrations of H_2O_2 polysulphoxides are formed

while in the second, slower stage, polysulphones are produced

Hitherto, the reactions of heterochain polymers have found practical use only when the side groups are considerably more reactive than the atoms in the macromolecular chain. Modification of the properties of cellulose is a classical example, which involves the esterification of the hydroxyl groups by inorganic or organic acids or the etherification of alkali salts of cellulose with alkyl chlorides.

From the historical viewpoint, the esters of nitric acid are the most significant derivatives of cellulose. They are produced industrially from cellulose and a mixture of nitric and sulphuric acid at 20 °C

$$+ 3n\,HNO_3 \longrightarrow \quad\quad + 3n\,H_2O$$

The optimum properties of cellulose nitrates are achieved with a nitrogen content of 11—13.5 % by weight; the complete substitution of each hydroxyl group by an NO_2 group yields a nitrogen content of 14.1 %.

Cellulose esters derived from acetic acid are produced in several types differing in their degree of substitution. The esterification of cellulose is conducted with a mixture of acetic anhydride and sulphuric acid, yielding the triacetate of cellulose; this is then successively hydrolyzed to the required ratio of acetate groups.

Cellulose may be dissolved in a mixture of NaOH and CS_2

to give the soluble xanthogenate of cellulose. The reaction is used to increase temporarily the solubility of cellulose in the main production of viscose fibres and cellophane. After processing of the viscose solution of the modified cellulose, the xanthogenate groups are removed by an acid medium and the original cellulose is regenerated.

Methyl cellulose, carboxymethyl cellulose and hydroxyethyl cellulose are the most important ethers of cellulose. They are prepared by reaction of methyl chloride, sodium chloroacetate and ethylene oxide with alkali salts of cellulose. The water-soluble ethers have on average one hydroxyl group per structural unit etherified. The most acidic hydrogen of the hydroxyl groups is that on the C-2 atom. The hydroxyl group of C-6 is sterically the most accessible but its reactivity has an intermediate value.

Such substitution reactions have found little application with other biopolymers, most of which have a more complex structure. Many chemically different mers are involved in the main chain, and attempted substitution reactions with a particular reagent may lead to changes of only some groups; even so the resulting impact on biological activity may be immense. The modification of proteins by formaldehyde is one of the least specific reactions, where in neutral or in a mildly alkaline medium, the hydrogen atoms of pendant amino groups of mers of glutamic and aspargic acids, asparagine and lysine are all substituted by methylol groups

$$P—\square—NH_2 + HCHO \cdot H_2O \quad \rightarrow \quad P—\square—NHCH_2OH + H_2O$$

N-methylol derivatives of proteins are less sensitive to biological attack.

As regards synthetic heterochain polymers, the aliphatic polyamides may be modified; thus hydrogen on a polyamide nitrogen atom may be substituted by chlorine or by an alkyl group.

A common drawback of reactions on structural units with several potential reaction sites of approximately equal reactivity lies in insufficient selectivity; another undesirable feature is the simultaneous decrease in the molecular mass of the modified polymer. With regard to these difficulties and to the wide choice of monomers for the direct synthesis of heterochain polycondensates, the possible modifications of these polymers have not been widely examined.

In summarizing the chemical transformations of the mers of macromolecular compounds we should be aware of the two ultimate aims of such procedures. The first is the synthesis of a material with novel or at least improved properties as compared with the parent polymer. Such an approach includes the reactions of relatively stable functional groups on a polymer with reactive low-molecular-mass compounds to give a resulting polymer with chemically stable functional groups. A different strategy may be seen in the functionalization of polymers, where from the original, less reactive groups are formed more reactive groups; some reaction sites may be transformed to centres of specific reactivity. One essential idea governs the study of such reactions, namely the synthesis and complete characterization of systems where the macromolecules play a role in addition to those of construction and support material. This will be dealt with in later chapters.

References

1. LENZ, R. W.: Organic Chemistry of Synthetic High Polymers. Interscience, New York 1967. KORSHAK, V. V.: Synthesis of Polymers by Modification Methods. Russ. Chem. Rev. *49*, 1135—1148, 1980.
2. KOLINSKÝ, M., DOSKOČILOVÁ, D., SCHNEIDER, B., ŠTOKR, J., DRAHORADOVÁ, E., KUŠKA, V.: Structure of Chlorinated Poly(vinyl chloride). Determination of the Mechanism of Chlorination from Infrared and NMR Spectra. J. Polym. Sci., A 1, *9*, 791—800, 1971.

3. QUENUM, B. M., BERTICAT, P., DE LA PENA, J. L., MILLAN, J.: Etude de la Chloration du Polychlorure de Vinyl en Fonction de sa Tacticite. Europ. Pol. J., *10*, 157—161, 1974.
4. PLATE, N. A.: Some Problems in the Reactivity of Macromolecules, in: Reactions on Polymers. MOORE, J. A., (Editor), Riedel, D., Dordrecht 1973, p. 169—229.
5. KENNEDY, J. P.: General Discussion on Chemical Modification. J. Macromol. Sci. Chem. A *12*, 327—342, 1978.
6. GOLUB, M. A.: Thermal Rearrangement of Polybutadienes with Different Vinyl Contents. J. Polym. Sci., *19*, 1073—1083, 1981.
7. PINAZZI, C., BROSSE, J. C., PLEURDEAU, A., REYX, D.: Recent Developments in Chemical Modification of Polydienes. Appl. Polymer Symposium, *26*, 73—98, 1975.
8. GELLING, I. R.: Epoxidized Natural Rubber in PVC-Rubber Composites, NR Technology, *16*, 1—2, 1985.
9. HRDLOVIČ, P., LUKÁČ, I.: Photochemically Initiated Maleic Anhydride Addition on Polystyrene Sensitized by Free and Bonded Sensitizer. J. Polym. Sci., Symposium *47*, 319—328, 1974.
10. MACKNIGHT, W. J., EARNEST, T. R.: The Structure and Properties of Ionomers. J. Polym. Sci. Macromol. Rev. *16*, 41—122, 1981.
11. CHEN, T. C. S., BUTLER, G. B.: Chemical Reactions on Polymers. III. Modification of Diene Polymers via the Ene Reactions with 4-Substituted 1, 2, 4-Triazoline-3, 5-Diones. J. Macromol. Chem., A *16*, 3, 757—768, 1981.
12. CAMBERLIN, Y., PASCAULT, J. P., RAZZOUK, H., CHERADAME, H.: Mise en Evidence du Caractere Sequence de l'Hydrogenation de Polydienes par Coupure Selective des Double Liaisons Makromol. Chem., Rapid Commun., *2*, 323—327, 1981.
13. CRIEGEE, R.: Mechanismus der Ozonolyse, Angew. Chem., *87*, 765—771, 1975.
14. RAZUMOVSKII, S. D., ZAIKOV, G. E.: The Effect of Ozone on Saturated Polymers (in Russian), Vysok. Soed. *24* A, 2019—2035, 1982.
15. KEFELI, A. A., VINICKAYA, Ye. A., MARKIN, V. S., RAZUMOVSKII, S. D., GURVICH, Ya. A., LIPKIN, A. M., NEVEROV, A. N.: Kinetics of Ozone Reaction with Rubbers (in Russian), Vysok. Soed., *19*, A, 2633—2636, 1977.
16. GENT, A. N., RODGERS, W. R.: Mechanochemical Reaction of Elastomers with Metals, J. Polym. Sci. Chem. Ed. *23*, 829—841, 1985.
17. BAUGHMAN, R. H., BREDAS, J. L., CHANCE, R. R., ELSENBAUMER, R. L. SHACKLETTE, L. W.: Structural Basis for Semiconducting and Metallic Polymer /Dopant Systems, Chem. Rev., *82*, 209—222, 1982.
18. BREDAS, J. L., CHANCE, R. R., BAUGHMAN, R. H., SILBEY, R.: Nonempirical Studies of the Electronic Properties of Highly Conducting Polymers. Int. J. Quantum Chem., Quantum Chem. Symp. *15*, 231—241, 1981.
19. FARRINGTON, G. C., HUQ, R.: Polyacetylene Electrodes for Non-Aqueous Lithium Batteries, J. Power Sources, *14*, 3—9, 1985.
20. SHACKLETTE, L. W., ELSENBAUMER, R. L., CHANCE, R. R., SOWA, J. M., IVORY, D. M., MILLER, G. G., BAUGHMAN, R. H.: Electrochemical Doping of Poly-*p*-phenylene with Application to Organic Batteries, J. Chem. Soc., Chem. Commun., 361—362, 1982.
21. WEGNER, G.: A Survey on Structure and Properties of Polymers with Metal-like Conductivity, Macromol., Chem., Macromol. Symp. *1*, 151—171, 1986.
22. POHL, H. A.: The Nature and Properties of Giant-Orbital Polymers J. Polym. Sci. A, Chem. Ed., *24*, 3057—3075, 1986.
23. PADIAS, A. B., HALL Jr., H. K.: Semiconducting Polymers via the High Temperature Free Radical Polymerization of Multinitriles J. Polym. Sci. A, Chem. Ed., *24*, 1675—1683, 1986.
24. DOOLEY, K. M., GATES, B. C.: Superacid Polymers from Sulfonated Polystyrene-Divinilbenzene: Preparation and Characterization. J. Polym. Sci. A, Chem. Ed., *22*, 2859—2870, 1984.

78

25. HALGAŠ, J., TOMA, S.: Polymers as Substrate Carriers in Organic Synthesis (in Slovak), Chem. listy, 77, 949—970, 1983.

26. OKAMOTO, J.: Radiation Synthesis of Functional Polymer, Int. J. Radiat. Appl. Instrum., Part C, Radiat. Phys. Chem., 29, 469—475, 1987.

27. SHERRINGTON, D. C.: The Effect of Polymer Structure on the Reactivity of Bound Functional Groups. Nouveau J. Chim., 6, 661—672, 1982.

28. WEBER, L.: Functionalization of Living Polymers, Results and Problems. Macromol. Chem., Macromol. Symp. 1, 317—329, 1986.

29. FRECHET, J. M. J., AMARATUNGA, W., HALGAŠ, J.: New Functional Polymers: Their Preparation and their Applications as Regenerable Reagents, Separation Media, or in Asymetric Synthesis. Nouveau J. Chim., 6, 609—616, 1982.

30. HARWOOD, H. J., CHEN, T. K., DVOŘÁK, A., GERKIN, T. A., KINSTLE, J. F., SHEPHERD, L.: Epimerization of Stereoregular Polymers. IUPAC 28 th Macromolecular Symposium, Amherst, Massachusetts July 12—16 1982, p. 97—99.

31. VEREANTEREN, F. F., DONNERS, W. A. B.: A ^{13}C Nuclear Magnetic Resonance Study of the Microstructure of Polyvinyl Alcohol. Polymer, 27, 993—998, 1986.

32. FEIT, E. D.: Photoresists: Photoformation of Relief Images in Polymeric Films, in: UV Curing. Science and Technology, Ed. PAPPAS, S. P., Stamford 1978.

33. KAEMPF, G., Special Polymers for Data Memories, Polym. J., 19, 257—268, 1987.

34. RAGHAVENDRACHAR, P., CHANDA, M.: Neighbouring Group Effect on the Kinetics of Acetalization of Poly(vinyl alcohol). Europ. Polym. J., 19, 391—397, 1983.

35. REMP, P.: Recent Results on Chemical Modification of Polymers. Pure Appl. Chem., 46, 9—17, 1976.

36. AKUMOTO, T.: Reactive Polymers, in: Speciality Polymers (translated into Russian). ISE, N., TABUSI, J., (Editors). Publishing House Mir, Moscow 1983.

37. CAIS, R. C., SPENCER, C. P.: The Dechlorination of Poly(vinyl chloride) by Zinc and Tributyl Tin Hydride. Europ. Polym J. 18, 189—198, 1982.

38. SUZUKI, T.: Chemical Modification of PVC. Pure and Appl. Chem., 49, 539—567, 1977.

39. ZUSSMAN, M. P., TIRRELL, D. A.: Backbone — Assisted Reactions of Polymers. J. Polym. Sci., 21, 1417—1422, 1983.

40. BELLUŠ, D., MAŇÁSEK, Z., HRDLOVIČ, P., SLÁMA, P.: Fries Rearrangement of Polyphenyl Esters to Polyhydroxyphenones. J. Polym. Sci., C, 267—277, 1967.

41. MARCO, C., BELLO, A., PEZENA, J. M., FATOU, J. G.: Oxidation of Polyhexamethylene Sulfide Single Crystals. Macromolecules, 16, 95—99, 1983.

IV. BRANCHING OF MACROMOLECULES

Macromolecules of branched structure originate from two processes. They may be synthesized either from polyfunctional monomers or from a linear polymer by its modification. The architecture of macromolecules formed in polyreactions depends mainly on the functionality of the monomers. Linear polymers are formed from bifunctional monomers which may bind two other structural units at most and grow in one dimension only. At any higher functionality of the monomer, the macromolecule starts to branch and crosslink. From the bifunctional and polyfunctional monomers, macromolecules of different degrees of branching may be synthesized, the length and distribution of branches being governed by statistical laws (*Fig. 4.1*). In principle, similar branching may also be obtained by modification reactions of linear macromolecules.

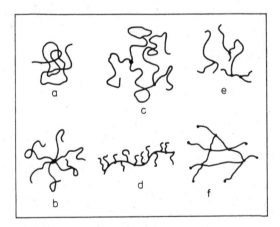

Fig. 4.1. Schematic illustrations of linear polymer molecules in (*a*) a coil, (*b*) a star, (*c*) a branched, (*d*) a comb-like, (*e*) a branched cascade, (*f*) a crosslinked structure of a macromolecule. The full circles denote the branching sites of the polymer chain

A. BRANCHED POLYMERS

These are produced not only in some chain reactions involving reactive by-products, but also from bifunctional monomers [1] as side-products. The long branches appearing during the radical polymerization of ethylene, vinyl acetate, methyl methacrylate and other bifunctional monomers are due to side-reactions of initially-produced polymer molecules $P^1 A P^2$. The transfer reaction of a free radical $\dot{R}$

$$P^1 A P^2 + R^{\cdot} \quad \rightarrow \quad P^1 \dot{B} P^2 + RH$$

activates the potential functionality in the structural unit of the polymer by inducing a radical centre inside the chain, which then reacts with other molecules of the bifunctional monomer M

$$P^1 \dot{B} P^2 + n M \quad \rightarrow \quad P^1 P^2 \dot{D} P^3$$

and a branched macroradical gradually appears. Its deactivation leads to the branched macromolecule. The degree of branching depends on the mechanism of decay of the branched macroradical. A single linear branch on a macromolecule is the result of either disproportionation or a transfer reaction, while two linear branches or one crosslinked chain are formed on the recombination of two macroradicals.

In the mechanism of branching, an important role is played by the copolymerization of a growing macroradical with the $C=C$ bonds of a pre-existing macromolecule. Each addition reaction of a macroradical creates two side branches if the double bond is non-terminal and one branch if the double bond is the terminal structural unit of the polymer backbone. Terminal double bonds in a polymer are the result of disproportionation and fragmentation of macroradicals and sometimes of the initiation of the growth reaction by a monomer radical.

The formation of long branches is usually an undesirable process in the preparation of a polymer since it increases the polydispersity and worsens the thermal processing of the polymer. In some polymers such as poly(vinyl chloride) [2], the branching sites decrease the thermal stability of the polymer and initiate its chain destruction.

The branching process is used in the synthesis of *"star"polymers* by anionic polymerization. Living chains of relatively low molecular mass ($2—3 \times 10^3 \text{g mol}^{-1}$) are terminated by reactive groups, which by a subsequent reaction, may give a very uniform branched structure. Such branched oligomeric rubbers are very easily moulded and after crosslinking, they can exhibit superior materials properties to classical rubbers [3].

Solutions of branched macromolecules are less viscous than those of linear polymers of the same molecular mass. This property apparently predetermined the character of the cascade-like branched polysaccharides, such as glycogen in animals and amylopectin [4] in plants tissue, as transportable reservoirs of energy. The main structural unit in molecules of glycogen and amylopectin is the glucopyranose ring

$$\begin{array}{c}
_6CH_2OH \\
\text{glucopyranose ring structure}
\end{array}$$

where the hydroxyl groups of C-6 are potential branching sites.

The *comb-like branching* of polymers is the result of isomerization of macroradicals. The macroradicals of poly(vinyl chloride) [4] and polyethylene [5, 6] isomerize

$$\text{—ĊXCH}_2\text{CHXCH}_2\text{CH}_2\text{X}$$

after approximately every 50 steps which leads to a higher ratio of short branches. Comb-like polymers of regular structure are prepared preferably either by the condensation reactions of reactive macromolecules with low-molecular compounds or by the polymerization of monomers having pre-existing side branches.

B. GRAFT COPOLYMERS

The two-step synthesis of branched macromolecules enables us to realize a new principle in the preparation of macromolecules, namely the attachment of side groups with a structure [7] different from that of the parent polymer. Such a polymer is designated a "grafted" copolymer. Clearly useful combinations of the properties of different polymers can be achieved in this way. A similar effect may be observed on mixing different polymers but a homogeneous dispersion is then attained only for particular pairs of macromolecular compounds. In polymer blends, the individual polymeric components usually form several separated phases. However polymer branches simulating the second component of such a mixture shift the incompatibility of the differing polymer chains to the

82

submicroscopic level, and from the macroscopic viewpoint, these branched copolymers appear to be homogeneous.

In theory, the simplest procedure for the preparation of graft copolymers is the condensation reaction of the proposed polymer branches with reactive functional groups of the original macromolecule. This is, however, difficult to conduct experimentally because of the repulsion of the incompatible macromolecular chains. A superior approach involves activation of the macromolecule by some initiator, branches being formed subsequently from the successive addition of monomers.

The choice of the method of generation of the active centres depends on the structure of the starting macromolecule and on the character of the monomer used. Radical grafting is the commonest approach being suitable for many combinations of polymers and monomers. Since mechanical mixing of the macromolecular reactants is usually accompanied by the appearance of free radicals (mechano-radicals) arising from the cleavage of mechanically strained chains, the grafting may occur even during the act of processing. It is not therefore surprising that the more reactive macromolecules are more suitable subjects for grafting. The practical applications of grafting are associated especially with amorphous polydienes and unsaturated polyesters.

The double bonds in the main chain of the polycondensation products of maleic anhydride and glycols are reactive centres which copolymerize with macroradicals of styrene and methyl methacrylate. The copolymerization is not completed at the stage of branching but continues until the polyester is wholly crosslinked.

In 1,4-polydienes, the vinylene double bonds are more numerous but less reactive than the vinyl side-groups present as a result of competitive 1,2-addition. Relatively low concentrations of very reactive anomalous units enable one to determine conditions of grafting which give branched copolymers without a high degree of crosslinking.

The initiation of the copolymerization of polydiene may also be induced by the abstraction of a hydrogen atom in the neighbourhood of a double bond; allylic hydrogens are transferred either to initiator radicals or to polymerizing macroradicals. Polydiene radicals then react with fresh monomer. In practice, the grafting may be performed as bulk, suspension or emulsion copolymerization. In the preparation of high-impact polystyrene, butadiene-styrene copolymer is dissolved in styrene and the solution is polymerized in a reactor equipped with a powerful screw mixer which transports the molten grafted polymer into the granulation machine.

The technically very important group of acrylonitrile — butadiene-styrene copolymers [8] is produced by two procedures.

a) by suspension polymerization of styrene with dissolved butadiene — acrylonitrile rubber,
b) by emulsion copolymerization of styrene and acrylonitrile occuring in the latex particles of polybutadiene rubber.

Methyl methacrylate can also be polymerized in the latex of natural rubber and, depending on the amount of polymerized methyl methacrylate, copolymers of different properties may be prepared.

The grafting process usually yields a nonhomogeneously grafted material. The heterogeneity of the product material is enhanced by the parallel formation of a homopolymer. Homopolymerization of the monomer is initiated by low-molecular-mass radicals derived from initiator, monomer or from macroradicals which have undergone fragmentation. The ratio of grafted and overall quantity of polymerized monomer is called the grafting efficiency. It depends on the reactivity of the grafted macromolecule, on the initiator used, on the concentration ratio of polymer to monomer as well as on the temperature, the solubility of the various components of the reacting system and many other factors. When the solubility of the monomer in the parent polymer exceeds that in the grafted polymer, then the efficiency of grafting decreases with increasing conversion. The grafting efficiency may be increased by selective generation of macroradicals on the polymer chain; the different rates of radical scission of some polymers and the reagent monomers brought about by ionizing radiation or through reactions of unstable groups present in a modified polymer may be used for this purpose.

a. Macromolecular Initiators

Radical polymerization is usually initiated by peroxidic compounds which are homolytically cleaved at their O—O bond. The resulting radicals react with monomers and start the growth of macroradicals. If the peroxy group is attached to a macromolecular chain then its decomposition can be used for direct activation of a polymer. Introduction of peroxy groups can be effected in various ways, e.g. reaction of a pendant acyl chloride with *tert*-butyl hydroperoxide

$$-CH_2CH- + (CH_3)_3COOH \longrightarrow -CH_2CH-$$
$$\underset{COCl}{\big|} \qquad\qquad\qquad\qquad \underset{COOOC(CH_3)_3}{\big|}$$

The macroradicals formed by thermal decomposition of such a polymer peroxide initiate grafting, while the *tert*-butoxy radicals also produced start ordinary homopolymerization. Although the efficiency of grafting for a macromolecular initiator is higher than achieved with a low-molecular peroxide, monomer and polymer system, it is still not the ideal solution.

The competitive formation of a homopolymer may be considerably suppressed by utilising as the initiation the redox reaction of hydroperoxide and transition metal ions

$$\overset{|}{\underset{OOH}{\overset{}{\rule{2cm}{0.4pt}}}} + Fe^{2+} \longrightarrow \overset{|}{\underset{O\cdot}{\overset{}{\rule{2cm}{0.4pt}}}} + Fe^{3+} + HO^-$$

or by exploiting other redox reactions of appropriate functional groups

$$\overset{|}{\underset{OH}{\overset{}{\rule{2cm}{0.4pt}}}} + Ce^{4+} \longrightarrow \overset{|}{\underset{O\cdot}{\overset{}{\rule{2cm}{0.4pt}}}} + Ce^{3+} + H^+$$

which generate the macroradical of a grafted polymer. Oxidation of hydroxyl groups by Ce(IV) salts may be applied to poly(vinyl alcohol), cellulose and starch. It is of interest that in this mode of activation of polysaccharides, the glucopyranose ring is cleaved

Many different methods have been attempted to achieve the direct chemical generation of radicals in macromolecules of different structures but these have been confined to the laboratory scale.

b. Radiation Grafting

The effort devoted to the optimization of grafting and to improving its efficiency led to the use of ionizing radiation which is particularly suitable for the less reactive saturated thermoplasts. Even though irradiation of a mixture of macromolecules and polymerizable monomer also produces unwanted homopolymer, the efficiency of grafting may be enhanced by a careful choice of monomer and polymer (*Table 4.1*). The most efficient grafting involves monomers which give a low radiation yield of radicals as compared with the grafted polymer. When poly(vinyl chloride) swollen with styrene is irradiated grafting occurs with an efficiency of almost unity.

If the polymer needs to be grafted with a monomer giving a high radiation yield of free radicals then the monomer is best supplied as a gas. This considerably lowers the absorption of radiation and increases the efficiency of grafting. Another approach involves the addition of a small amount of polyfunctional

Monomer	G	Polymer	G
Styrene	0.7	Polystyrene	2
Butadiene	1	Cis-1,4-polyisoprene	3
Acrylonitrile	5	Polyethylene	7
Methyl methacrylate	8	Poly(methyl methacrylate)	9
Vinyl acetate	9	Poly(vinyl acetate)	10
Vinyl chloride	11	Poly(vinyl chloride)	12

monomer which mediates the attachment of homopolymer radicals to the grafted substrate.

When the radiation-chemical yields of free radicals from the monomer and polymer are particularly adverse, the efficiency of grafting may be increased by an inhibitor soluble in the monomer phase where it inhibits the chain polymerization. This is particularly suitable for the surface grafting of finished products (fibres, foils).

Since heterogeneous grafting depends on the mutual solubility of the reacting components and on the rate of diffusion of monomer into the polymer, diluents and swelling agents may affect it significantly (*Fig. 4.2*) [9]. On slight swelling of the polymer, the grafting zone gradually moves from the surface into the bulk, and the reaction is accelerated accordingly.

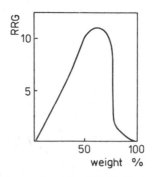

Fig. 4.2. Plot of the relative initial rate of grafting (RRG) of a polytetrafluorethylene film versus the concentration of *N*-vinylpyrrolidone in benzene (in w. %)

The overall radiation dose, which determines the average number of radicals formed on the macromolecular backbone, changes the average number of grafted branches along the chain while the intensity of radiation, which determines the instantaneous concentration of radicals in the system, controls the length of the branches. The degree of polymerization of the grafted branches is also affected by transfer reactions of monomer, polymer and solvent.

One way of increasing the grafting efficiency is to irradiate the polymer alone. When monomer is added subsequently, the fraction of macroradicals stabilized in the polymer which initiate grafting increases and it may become the dominant process. However, since some macroradicals decay during irradiation of the polymer, the efficiency of grafting related to the radiation dose is lower.

This technique may also be accompanied either by reduction of the molecular mass of the polymer or, in an inert atmosphere, by crosslinking of grafted macromolecules. In the presence of air, alkyl macroradicals react with oxygen molecules; the peroxyl radicals formed abstract hydrogen from surrounding molecules and hydroperoxyl groups are formed on the polymer chain. As indicated above, these thermally unstable groups may be used for the chemical initiation of grafting.

An interesting variant of radiation grafting is based on irradiation of vinyl, vinylidene and diene monomers adsorbed on inorganic compounds such as SiO_2, TiO_2, $CaCO_3$, Al, etc. The chemical fixation of a polymer layer on the surface of these materials alters their properties and enables the synthesis of composite materials [10]. The grafting is initiated here by the surface radicals or by ionic sites [11]. The grafting of macromolecules onto inorganic particles may also be effected by other methods such as the milling of inorganic materials in the presence of monomers and by chemical activation.

c. Nonradical Grafting

The choice of the growth polyreaction for grafting onto the original polymer chains depends on the character of the functional groups present. Solutions of aromatic vinyl polymers give complexes with alkali metals which may initiate the anionic polymerization of monomers. At initiation, a chemical bond between the functional group of, say, poly(vinyl naphthenate) and a growing branch is formed. Similarly, the iodophenyl group in polystyrene, after its reaction with butyl-lithium, is transformed into a lithiumphenyl complex which may initiate the anionic polymerization of acrylonitrile

Grafting by an anionic mechanism is highly efficient and can be used to regulate the number and length of the branches. The drawbacks of the reaction lie in the stringent requirements of the purity of the reactants and in the relatively few systems amenable to grafting by the anionic mechanism. Addition of the ion-pair consisting of oligomeric ($n \sim 20$) polyisopropenyl-lithium to chloromethylated polystyrene

$$C_4H_9(C_5H_9)_{\bar{n}}Li^+ + ClCH_2\text{—}\bigcirc\text{—}CH \longrightarrow C_4H_9(C_5H_9)_n\text{—}CH_2\text{—}\bigcirc\text{—}CH + LiCl$$

is an example of a nonchain reaction giving comb-like grafted copolymers.

As regards the grafting of chlorine-containing polymers, the cationic polymerization of a limited number of monomers may be used

$$-CH_2CH- \xrightarrow{AlCl_3} -CH_2\overset{+}{C}H- \xrightarrow{n\,CH_2=CHC_6H_5} -CH_2CH-$$
$$\underset{Cl}{|} \qquad\qquad \underset{Al\bar{C}l_4}{|} \qquad\qquad (\overset{|}{C}H_2CHC_6H_5)_n{}^{+-}AlCl_4$$

Unfortunately, the reaction is accompanied by extensive transfer and fragmentation and the grafting efficiency is very low.

Another, less studied possibility for grafting is the catalyzed dismutation (metathesis) of alkenes [12]

$$\begin{matrix} R^1CH=CHR^2 \\ + \\ R^3CH=CHR^4 \end{matrix} \rightleftharpoons \begin{matrix} R^1CH \\ \| \\ R^3CH \end{matrix} + \begin{matrix} CHR^2 \\ \| \\ CHR^4 \end{matrix}$$

which occurs very rapidly even at room temperature in the presence of heavy elements of the fifth and seventh groups of the periodic table, particularly compounds of W, Mo and Re in combination with those of Al and I.

The grafted copolymers may be formed via the reaction of a double bond on the macromolecular backbone with that of some other macromolecule. Polymers with an unreactive backbone can be used for grafting provided the reactive side (alkene) groups are present through incorporation by copolymerization during the original preparation. In the case of terpolymers of ethylene-propylene rubbers, the reactive component is the third bifunctional comonomer, which copolymerizes with one double bond of the macromolecular backbone, while the second double bond reacts with polyalkene

Those compounds having isolated double bonds are the most suitable for metathesis, conjugated double or triple bonds usually deactivating the catalyst. The catalyst may also be poisoned by water, acids, alcohols, amines, esters, oxygen and many other compounds. Apart from deactivation of the catalyst, metathetic grafting is impeded by phase separation of dissimilar macromolecules, and from this viewpoint, oligomeric reactants are more suitable.

The main current research effort is directed towards new grafting procedures and novel polymer substrates [13].

The theoretical treatment of polymer grafting encounters problems in the characterization of reaction products [14]. The reaction system is a mixture of the original homopolymer and molecules of the copolymer grafted to varying degrees. Their analytical separation is very difficult mainly due to the presence of the grafted copolymer which makes the physical properties of the originally unlike polymers more uniform. The determination of the structural parameters of isolated copolymers, such as the number and length of the branches or their distribution along the polymer chain requires selective destruction techniques, which are experimentally very difficult to realise and may yield ambiguous results.

References

1. WOLF, CH., BURCHARD, W.: Branching in Free Radical Polymerization due to Chain Transfer, Makromol. Chem. *177*, 2519—2538, 1976.
2. HJERTBERG, T., SORVIK, E. M.: Formation of Anomalous Structures in PVC and their Influence on the Thermal Stability.
   2. Branch Structures and Tertiary Chlorine. Polymer, *24*, 673—684, 1983.
3. DOLGOPLOSK, B. A.: On Some New General Approaches to Rubber and Resin Synthesis (in Russian). Dokl. Akad. Nauk SSSR *263*, 1142—1143, 1982.
4. SCHRODER, E., BYRDY, M.: Studien zur Verzweigung von Kettenmoleculen, Faserforschung und Textiltechnik, *28*, 229—233, 1977.
5. MATTICE, W.: Complex Branch Formation in Low-Density Polyethylene, Macromolecules, *16*, 487—490, 1983.
   HANNEY, M. A., JOHNSTON, D. W., CLAMPITT, B. H.: Investigation of Short Chain Branches in Polyethylene Pyrolysis, GCMS, Macromolecules, *16*, 1775—1783, 1983.
6. HAY, J. N., MILLS, P. J., OGUJANOVIC, R.: Analysis of ^{13}C FT-NMR Spectra of Low Density Polyethylene by the Simple Method, Polymer, 27, 677—680, 1986.
7. Block and Graft Copolymerization. *1*, 1973, *2*, 1976. CERESA, R. J., (editor), J. Wiley and Sons, London.
8. STARÝ, M., POHL, I., KAŠÍK, B.: Terpolymers of ABS, Processing and Use (in Czech), SNTL, Praque, 1977.
9. CHAPIRO, A.: Radiation Induced Grafting, Radiat. Phys. Chem., 9, 55—67, 1977.
10. IVANČEV, S. S., DMITRENKO, A. V.: The Way of Preparation of Filled Polymer Composites by Radical Polymerization, Russ. Chem. Rev., *51*, 670—683, 1982.

11. STEINIKE, U., LINKE, E.: Die Rolle von Anregungszustanden und Structurdefekten bei Mechanisch Aktivierten Festkorperreaktionen. Z. Chem., *22*, 397—401, 1982.
12. STRECK, R.: Die Olefin-Methathese, ein Vielseitiges Werkzeug der Petro und Polymerchemie. Chem. Zeit., *99*, 397—413, 1975.
13. SHIGENS, Y., KONDO, K., TAKEMOTO, K.: Radiation-Induced Graft Polymerization of Styrene onto Chitin and Chitosan, J. Macromol. Sci., Chem. A *17*, 571—583, 1982.
14. STANNET, V., MEMETEA, T.: Some Unsolved Problems Concerned with Radiation Grafting to Natural and Synthetic Polymers. J. Polym. Sci., Polym. Symp. *64*, 57—69, 1987.

V. LINKING AND CROSSLINKING OF MACROMOLECULES

The chemical linkage of essentially linear molecular fragments into assemblies, for which the term molecule loses its usual meaning, represents the most significant approach in the practical exploitation of polymer chemistry. It encompasses the preparation of phenol formaldehyde resins, originating at the beginning of this century, and several other contemporary thermosets such as melamine resins, epoxides, unsaturated polyesters and polyurethanes. The crosslinking of asphalt in the formation of the first photographic negatives (1815), the early invention of the vulcanization of natural rubber (1838) and particularly the recently developed understanding of the rules governing the crosslinking of elastomers and biopolymers also come into this category.

The linkage of two or three types of macromolecule solely through their end groups by a bifunctional bridge is the objective in the preparation of block copolymers. Depending on the chemical structure of the individual blocks, the resulting copolymer can, but need not have, the properties of a physically crosslinked polymer. Chemical crosslinking is achieved when the end groups are linked by a multifunctional (i.e. more than bifunctional) agent. This type of crosslinking is advantageous mainly because of the absence of loose chain ends which are elastically ineffective in the crosslinked polymer.

A. CROSSLINKING

Crosslinked polymers comprise about one third by volume of macromolecular materials currently produced. The variety of types and applications of crosslinking indicates the important impact of this structural modification on the properties of macromolecular systems. Crosslinked polymers are usually formed from low-molecular multifunctional monomers. However, the volume of polymers additionally crosslinked after their preparation is considerable, and amounts worldwide to millions of tons of materials annually. The former method of preparation of crosslinked polymer is typical of resins, the latter of elastomers.

a. Types of Crosslinks

Several possibilities for macromolecular linkage result from the diversity of reactions leading to a new chemical bond. To facilitate the following account of the methods of crosslinking, we have chosen the chemical composition of the crosslink unit as the criterion of classification.

Crosslinking as a means of fixing a particular structure of a macromolecular system is employed not only by man in modification of polymer properties. In Nature the intermacromolecular crosslinks act as stabilizers of optimal arrangements of the functional parts of an organism. For example, macromolecules are bridged by short hydrocarbon bonds and by disulphide bonds. Crosslinking by noncovalent bonds (hydrogen bonds, hydrophobic interactions) occurs frequently in biopolymers such as DNA. The latter, rather unstable chain linkage, permits a facile rearrangement of the existing structure, mediating thereby important changes in the function of living matter.

Noncovalent crosslinks are also used in synthetic polymers, for example in thermoplastic rubbers based on ABA copolymers as well as in ionomers.

1. Linkage via Carbon-Carbon Bonds

Formation of C—C covalent bonds between molecules is based on the elimination of side atoms or groups and on the subsequent linking of the remaining parts of the molecules according to the following scheme

$$
\begin{array}{c}
R^1CHR^2 \\
| \\
X^1 \\
| \\
X^2 \\
| \\
R^3CHR^4
\end{array}
\longrightarrow
\begin{array}{c}
R^1CHR^2 \\
| \\
R^3CHR^4
\end{array}
+ X^1{-}X^2
$$

where X is H, Cl, F, CH_3, etc. The several variants of this process enabled carbon-carbon crosslinking to be investigated most thoroughly. This reaction is used in the technological modification of the properties of polyethylene. The elimination of the hydrogen atom is achieved here by ionizing irradiation or by small free radicals in the transfer reaction [1]

$$R^1CH_2R^2 + RO^{\cdot} \rightarrow ROH + R^1\dot{C}HR^2$$

The resulting macroradicals form the carbon-carbon crosslink via the recombination reaction

$$
R^1\dot{C}HR^2 + R^3\dot{C}HR^4 \longrightarrow
\begin{array}{c}
R^1CHR^2 \\
| \\
R^3CHR^4
\end{array}
$$

Elimination of low molecular products is also brought about by high temperatures, and crosslinked, carbonized residues are the final products of thermolysis of numerous polymers. *Unstable cyclic structures* may also serve as precursors of C—C crosslinks. This type of crosslinking proceeds during thermolysis of polymers with aromatic rings in the main chain, thus

Carbon-atom crosslinks can also be formed via *the polymerization mechanism* if the chain contains vinyl groups, thus

It should be pointed out, however, that double bonds in the macromolecular backbone react much more slowly by the polymerization mechanism. This fact may be connected with an unfavourable reactant orientation, because two such backbone groups achieve the correct alignment with greater difficulty than freely rotating side- or end-groups. The radical reaction of a double bond in the backbone proceeds at a sufficient rate if the macroradical can react with a polymerizable monomer. This fact is employed industrially for the crosslinking of unsaturated polyesters with styrene or methyl methacrylate. C—C bond crosslinks are also formed here but the linkage represents a homopolymer sequence.

From the theoretical viewpoint attention should be paid to the formation of intermacromolecular carbon-carbon crosslinks by *dismutation (metathesis)* of cyclo-olefinic side groups [2]

The linkage is again composed of several carbon atoms but in this case of a cyclic diolefinic structure there are two parallel links emanating from the same carbon atoms on the linked macrochains.

2. Crosslinking through Sulphur

A macromolecular network having crosslinks through sulphur is a characteristic of vulcanized polydienes and some proteins. The crosslinking of natural and synthetic rubber by sulphur (vulcanization) is of immense significance technically.

Elementary sulphur has a low efficiency for crosslinking in rubber. The formation of a single crosslink between two polydiene molecules, which is essential for the required modification in properties, requires from 40 to 50 atoms of sulphur. This is due to the prevalence of simultaneous side reactions such as intramolecular cyclization,

the formation of S—H bonds and the occurrence of polysulphide crosslink bridges as well as to their formation in pairs

The latter situation have the same effect on the physical properties as a single crosslink.

The number of monoatomic and diatomic sulphur linkages may be increased significantly by additives (vulcanization accelerators), such as 2-mercaptobenzothiazole, etc., when only about two sulphur atoms are necessary for one crosslink.

Monoatomic sulphur linkages may be introduced into the polymer directly from a suitable accelerator, e. g., from tetramethylthiuram disulphide or sulphur chloride

The latter reaction is used in the vulcanization of thin layers of natural rubber.

Besides sulphur and/or accelerator, zinc oxide, stearic acid, carbon black, inorganic fillers, and some other additives are used in the processing of poly-dienes. Depending on the structure of the parent polydiene, they may have additional effects on the formation of the macromolecular network as well as on the extent of side reactions. The complexity of vulcanization systems is, more-over, accentuated by the heterogeneous character of insoluble zinc oxide which is used as an activator of the process initiating polyaddition reactions. The detailed description and interpretation of the mechanism of vulcanization is accordingly very difficult, and at best only an empirical understanding has been achieved in many cases even today.

The complex group of additives changes its chemical character in the course of vulcanization and cannot be therefore considered as a classical catalyst. The main primary process is the reaction together of sulphur, accelerator and activator to give reaction intermediate which then participates in the vulcaniza-tion process [3]. This may be illustrated by a tentative mechanism for the vulcanization of rubber (RH) in the presence of benzothiazole (B) where the low-molecular mass polysulphide of benzothiazole is formed initially; the latter subsequently reacts with rubber

$$B - S(S)_x SB \longrightarrow B - S(S)_x^{\cdot} + {}^{\cdot}SB$$
$$RH + {}^{\cdot}S - B \longrightarrow B - SH + R^{\cdot}$$
$$R^{\cdot} + {}^{\cdot}S(S)_x - B \longrightarrow R - S(S)_x - B$$

The complex of Zn ions and acelerator binds elementary sulphur more easily, and the intermediate polysulphide attacks the allylic position of rubber

The pendant polysulphide groups attached to the polydiene initiate further reaction with another macromolecule and form the corresponding linkage.

Polysulphides are known to be adsorbed on the polymer surface. Zinc oxide may, therefore, alternatively participate in the direct formation of crosslinks, substituting the polysulphide linkages by the polar bonds of some groups in the accelerator attached to the polydiene and its surface. Another possibility is the heterogeneous condensation of ZnO with thiol groups.

$$2\,R{-}SH + ZnO \longrightarrow RSZnSR + H_2O$$

The technical product of rubber vulcanization has a large variety of crosslinks. The formation of mono and diatomic sulphur bridges requires less sulphur in the formulation of the process. The other crosslinks, however, give us the possibility of optimising the properties of the product for a particular application.

Sulphur and its compounds may also be used as reagents in the crosslinking of some saturated polymers such as polypropylene, copolymers of propylene, etc. [4]. The alkyl macroradicals formed from these polymers exhibit a greater tendency to fragmentation than to recombination. Addition of such a radical to sulphur decreases the probability of its fragmentation and increases the number of mutual recombinations. Alternatively, the crosslinking may proceed via intermediately formed double bonds and the mechanism may be similar to that of rubber vulcanization.

Disulphide bridges are important structural units in some proteins and other biological macromolecular systems. Intramolecular or interchain disulphide bonds are formed in the mild oxidation of thiol groups on cysteine units in proteins.

$$
\begin{array}{c}
| \\
CO \\
| \\
2\ CHCH_2SH \\
| \\
NH \\
|
\end{array}
\quad
\underset{\text{reduction}}{\overset{\text{oxidation}}{\rightleftharpoons}}
\quad
\begin{array}{c}
| \qquad\qquad | \\
CHCH_2SSCH_2CH \\
| \qquad\qquad |
\end{array}
$$

Disulphide crosslinks contribute to the maintenance of the spatial organization of proteins and to the stabilization of its tertiary structure. The reverse reduction of the disulphide linkage leads to the destruction of the spatial structure. This mechanism is important in the function of some biological regulators and enzymes (insulin, ribonuclease). It has also found practical application in the permanent waving of hair. The reduced and noncrosslinked array of keratin macromolecules in hair is shaped to the waves which are fixed by oxidation and consequent crosslinking via disulphide bridges.

The principle of the reversible formation and destruction of the macromolecular network is also attractive in rubber processing since it would allow its multiple utilization. The idea has so far been exploited only in the case of thermoplastic rubbers where, however, physical crosslinking is taking place.

3. Heteroatom Linkages

The crosslinking of polymers with reactive functional groups may be accomplished by their reaction with bifunctional reagents which can link together two centres on different macromolecules. Chlorosulphonated polyethylene can thus be crosslinked, e.g. by benzidine

$$\text{PE}\left|-SO_2\!-\!Cl\;\;H\!-\!NH-\!\left\langle\bigcirc\right\rangle\!\!\left\langle\bigcirc\right\rangle\!-\!NH\!-\!H\;\;Cl\!-\!SO_2-\right|\text{PE}$$

Similarly, halogen-containing polymers and polymers with hydroxyl, carboxyl and other groups may also be more or less efficiently crosslinked in this way.

The set of reactions leading to the formation of a macromolecular network, especially the crosslinking of polysaccharides and casein is of industrial importance. It involves a two-step reaction of the hydroxyl groups in the polysaccharide and/or amino groups in casein with formaldehyde which results in methylene bridges.

In polysaccharide chemistry use is also made of their reaction with epichlorhydrin (1-chloro-2,3-epoxypropane) [5]. The crosslinking proceeds as a stepwise reaction of the hydroxyl groups of the polysaccharide (ROH)

$$ROH + CH_2\!-\!CHCH_2Cl \longrightarrow ROCH_2CHCH_2Cl \xrightarrow{\;NaOH\;} ROCH_2CH\!-\!CH_2$$

Oxirane rings on the polysaccharide react with the hydroxyl groups of another macromolecule, and a 2-hydroxypropyl ether linkage

$$ROCH_2CH(OH)CH_2OR$$

is thus formed. In the presence of water, the oxirane rings give 1,2-dihydroxypropyl ether polysaccharide, and crosslinking is slowed down.

Condensation reactions involved in crosslinking simultaneously produce low-molecular-mass compounds which may either detract from the materials properties of the product or reduce the rate of the process. In polyaddition reactions the polyfunctional crosslinking agent reacts with the reactive groups on the macromolecules without formation of low-molecular-mass products; this case may be represented by the polyaddition of borane to unsaturated polymers

$$BH_3 + \quad\quad\quad \longrightarrow \quad H\overset{H}{\underset{H}{B}}$$

On reaction of the crosslinking agent with macromolecular chains, the network structure formed contains loopholes of various dimensions. A more homogeneous network may be achieved when the reactive functional groups are regularly distributed along the macromolecular backbone. Crosslinking

97

mediated by polydimethylsiloxane end-groups and induced by alkyltriallyloxy-silane is an example. The regular dimensions of the loopholes in the network depend on the polydispersity of the polydimethylsiloxane. The crosslinking, the rate of which depends on the catalyst and its concentration, may occur even at ambient temperature. Crosslinking may be also performed with macromolecules which themselves contain several mutually reacting functional groups.

Crosslinking occurring via latent functional groups, which become active on reaction with atmospheric moisture finds increasing use in practice. It particularly concerns trichloro- and trialkoxysilyl groups attached to a macromolecular backbone. The crosslinking occurs as a result of hydrolysis and subsequent condensation, and siloxane crosslinks are formed

$$RSiCl_3 \xrightarrow{H_2O} RSi(OH)_3 \xrightarrow{-H_2O} R\underset{|}{\overset{|}{Si}}O\underset{|}{\overset{|}{Si}}R$$

Silyl functional groups may be incorporated into a macromolecule either in a growth polyreaction or in the supplementary reaction of vinyl trialkoxysilane with polymer molecules.

Condensation reactions of functional groups are used in Nature for cross-linking of proteins giving products which function in the supportive and protective tissues (skin, cartilage, bones, dentin) of higher organisms. The biosynthesis of collagen takes place in the ribosomes of cells. After their secretion into the microcellular area, the helical macromolecules of collagen are aggregated by electrostatic forces into more organized structures, such as fibrils, which are then fixed by crosslinking. Covalent crosslinks between macromolecules in fibrils bring about the integrity and stability of the collagen matrix in the tissue. Crosslinking increases the resistance of the matrix towards effects such as thermal denaturation; it also reduces the probability of enzymatic scission and leads to the appearance of elastic deformability and to the loss of solubility.

Most crosslinks are formed by the enzymatic oxidative deamination of lysine and hydroxylysine substituents on the aminoacid structural units of collagen molecules (cM) and by their subsequent reactions [6]. Aldehydic groups formed by oxidation enter a condensation reaction with the amino groups of lysine or hydroxylysine

$$\Big|-(CH_2)_2\underset{\underset{OH}{|}}{C}HCHO + H_2NCH_2\underset{\underset{OH}{|}}{C}H(CH_2)_2-\Big|$$

collagen
macromolecule
cM   cM

↓

$$\Big|-(CH_2)_2COCH_2NHCH_2\underset{\underset{OH}{|}}{C}H(CH_2)_2-\Big|$$

cM   cM

In elastin and other proteins, lysine, norleucine, tyrosine and alanine mers are thus linked together by the effect of oxidative enzymes [7]. The several kinds of crosslinks formed probably have different functions which may be associated with the different stabilities of crosslinks formed during the growth and regeneration of tissues in living organisms. The appearance of such a variety of crosslinks is necessary because of the different rates of crosslinking of macromolecules in particular phases of evolution of the organism.

The crosslinking of elastin occurs during the synthesis of the biomacromolecules. It is evidenced by the considerably higher ratio of desmosin (by 150 %)

in the elastin walls of the aorta of one-year old chickens when compared with those of an embryo of 12 days [8].

The crosslinking of collagen is accomplished technically by salts of Cr, Al and Zr or by such organic materials as *tanning compounds*, viz. aldehydes and quinones. In the tanning of leather, crosslinking is brought about by the ionized groups of collagen which coordinate with chromic ions. The mechanism of tanning by zirconium salts is similar but here the coordination bonds are mediated by the amino groups of the collagen chains (particularly the arginine moieties) where one Zr species binds at least two macromolecules. In the case of aluminium salts the crosslinking also occurs via the carboxyl groups of collagen but the coordination bonds are less stable than those with Cr. In tanning by compounds such as polyhydroxyphenols, hydrogen bonds are formed between the amidic nitrogen of the collagen chain and the hydroxyl groups of the polyphenol

Such interaction creates relatively easily transferable crosslinks. Stable covalent crosslinks mostly involve pendant amino-groups and the quinone units of

99

the oxidized part of the tanning compound. The structure of the crosslink bridge resembles that formed in the reaction of the lysine units of a protein with 1,4-benzoquinone

$$\text{CH(CH}_2)_4\text{NH} \overset{\displaystyle O}{\underset{\displaystyle O}{\bigcirc}} \text{NH(CH}_2)_4\text{CH}$$

As regards aldehydes, formaldehyde has the highest tanning efficiency, forming crosslinks with the amino and hydroxyl groups of collagen, as pointed out previously.

The crosslinking reaction of collagen has found the widest practical applications in the technology, processing and refinement of leather, pelts and gelatin [9].

Although not all the known procedures for effecting the covalent crosslinking of macromolecules are referred to here, one may envisage the numerous practical applications from the above examples given. Research effort nowadays is focused not merely on devising further crosslinking reactions but pays attention to the mutual relation of the procedure for crosslinking and its mechanism, and the homogeneity and structure of the network formed, etc.

Microheterogeneity in the distribution of crosslinks may sometimes result as a consequence of a nonhomogeneous dispersion of the crosslinking agent in the polymer or during the crosslinking process itself. At the point of mechanical mixing of the reactants, the homogeneity of the crosslinks subsequently formed depends on the solubility of the crosslinking agent in the polymer as well as on the rate of crosslinking when compared with the rate of dispersion of initiator. As soon as the crosslinking reaction begins, mixing should be stopped, lest the existing crosslinks are disrupted. The crosslinks diminish the mobility of the macromolecules and hence the diffusion of low molecular mass compounds in the polymer, an effect which contributes to the nonhomogeneous distribution of crosslinks. To ensure a homogeneous distribution of crosslinks in the polymer, the crosslinking should be performed with a homogeneous mixture of oligomers having reactive functional groups capable potentially of forming the crosslinked network [10].

4. Crosslinking by Secondary Bonds

Besides the mutual entanglements of long linear molecules, a reversibly crosslinked system may also be achieved with copolymers having 5—10 % mol of acidic functional groups attached to their chain. Carboxyl, sulphonic and other acid groups may be converted to salts of Zn, Mg or other suitable metal. The neutralized pendant groups aggregate to produce a separate phase (*Fig. 5.1*) and form there secondary bonds which crosslink the macromolecules. The crosslinked structure of such polymers (ionomers) may be broken down by heating. Reversible crosslinking is an equilibrium process which may be repeated after cooling down of the thermally moulded ionomer.

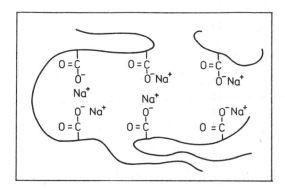

Fig. 5.1. Aggregation of polar functional groups in an ionomer

A similar type of crosslinking is also used for thermoplastic rubbers with a block copolymer structure; here, sequences of structural units with rubber-like properties are inserted, which are terminated with polystyrene blocks. The rubber segments form a continuum with islands of polystyrene aggregates (*Fig. 5.2*). At ambient temperatures the polystyrene segments are in the glassy

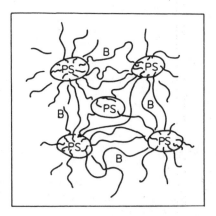

Fig. 5.2. Illustration of the arrangement of polydiene (B) and polystyrene (PS) blocks in the copolymer PS-B-PS. The aggregates of styrene based structural units in the glassy state form crosslinks with the physically crosslinked thermoplastic rubber

101

state and fix the flexible polybutadiene or polyisoprene parts of the macromolecule, thus physically crosslinking the material. By heating to above T_g of polystyrene, the properties of the crosslinked polymer disappear, but are regenerated on cooling. As expected, if the polystyrene segments are inside a block copolymer chain, then physical crosslinking is impossible, and equally inefficient are block copolymers of type A-B. The physical crosslinking of block copolymers is possible with a combination of crystallizable and amorphous segments.

The crosslinking of macromolecules can occur through electrostatic forces between their functional groups when polyelectrolytes of opposite charge [11] are mixed together. The crosslinking may be carried out by the mixing of aqueous solutions of a polybase and polyacid or by a matrix polymerization of the monomer on a polyelectrolyte of opposite charge. Due to the attractive interactions of opposite charges (*Fig. 5.3*), an insoluble complex of a polymer salt is formed which is heavily swollen by water. In the complex, the macromolecules, usually of different polyelectrolytes, are found in the approximate molar ratio 1 : 1 regardless of the initial concentrations of the respective components in water. Poly(4-vinyl pyridine), poly(vinyl amine), poly (L-lysine) and polyethyleneimine function as the polycation carriers while the polyanions are usually derived from polystyrenesulphonic acid, carboxymethylcellulose, cellulose xanthogenate, heparin, etc.

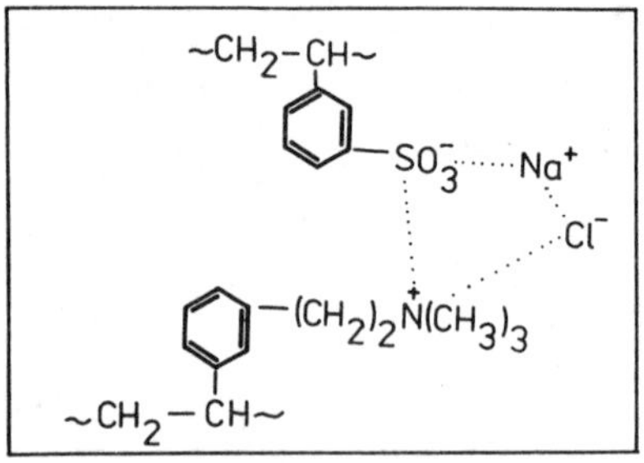

Fig. 5.3. An example of the electrostatic (ionic) linkage of macromolecules of a polyacid and a polybase

b. Methods of Crosslinking

The condensation reaction of functional groups of macromolecules may proceed directly or via polyfunctional monomers. Such crosslinking can be induced on adding the crosslinking agent to the polymer or simply on heating which induces the formation of covalent bonds between the reactive groups of the macromolecules.

Of significance also are procedures where chemically stable macromolecules are activated by irradiation or by small radicals. The transformation of inactive macromolecules into reactive macroradicals and the successive formation of crosslinks by the recombination of macroradicals is not only of practical use but contributes also to our knowledge of solid phase reactions.

1. Radiation Crosslinking

In the early days of radiation research it was suggested that the very high energy of γ-photons which is nearly 10^5 times higher than the dissociation energy of chemical bonds, might give rise to new chemical reactions. It was shown, however, that γ-rays provide only a new method of generating of radicals, ions, free electrons and excited states of functional groups. The presence of reactive particles in a polymer may be detected by various methods such as electron spin resonance (ESR), optical spectroscopy and thermoluminescence. The yields of these particles and the occurrence of their subsequent reactions depend only slightly on the kind of high energy radiation. The further course of reaction is not different from that found with the chemical generation of reactive intermediates. The advantage of radiation initiation lies in the possibility of generating reactive intermediates in the solid state over a large range of temperature and concentrations.

a. Polyethylene

The crosslinking of polyethylene is one of the most economically successful results of radiation chemistry research [12] comparable even with the use of ionizing radiation in the sterilization of foodstuffs and medical instruments and in tumour therapy and the mutagenesis of plants. The effect is associated obviously with the observation that a small perturbance to the structure of a macromolecule may induce large changes in its physical and biological properties.

The chemical consequences of irradiation are expressed by *the radiation yield G* which represents the number of molecular changes accomplished in a material by a dose of 100 eV.

For long-chain alkanes, the *G*-value for scission of the C—C bond is 2.7 while that for the C—H bond is 4.3, although C—H bonds are stronger than C—C bonds. Taking into account the large energy excess of the γ-photons, the bond dissociation energy plays no significant role here. The unexpected difference may be attributed to the easier dissipation of energy over several mutually linked C—C bonds when compared to the more isolated C—H bonds. Another reason lies in the greater mobility of abstracted hydrogen atoms which escape from the site of their generation more easily than the two heavier alkyl radicals formed in the dissociation of a C—C bond. The fast separation of radicals from the initially formed radical pair reduces the probability, of their reverse combination.

Polyethylene irradiated in vacuo gives predominantly secondary polyalkyl radicals

$$-CH_2\dot{C}HCH_2-$$

In the temperature interval from 50 to 250 K, neither the ESR spectrum of hydrogen atoms nor of primary alkyl radicals $-\dot{C}H_2$ can be detected. Any transient hydrogen atoms and primary alkyl radicals abstract hydrogen from surrounding macromolecules and convert the polymer to secondary radicals. Other types of less reactive radical may be observed at elevated temperatures or at higher radiation dose. Allyl radicals appear at ca. 300 K

$$-CH_2\dot{C}HCH=CHCH_2-$$

and at higher doses polyene radicals are found

$$-\dot{C}H(CH=CH)_nCH_2-$$

By recombination of secondary carbon radicals, crosslinks between macromolecular chains of polyethylene are formed with a radiation yield about 1 at temperatures below 230 K. The theoretical value of the radiation yield of crosslinking is about two times higher, assuming an absence of cleavage of C—C bonds and that all alkyl radicals recombine. The lower value found for the crosslinking efficiency is due to the destruction of the main chain and also to disproportionation of the secondary radicals

$$-CH_2\dot{C}HCH_2- \qquad -CH_2CH_2CH_2-$$
$$\longrightarrow$$
$$-CH_2\dot{C}HCH_2- \qquad -CH=CHCH_2-$$

which leads to the formation of some double bonds in the irradiated polymer. Some of these double bonds may also be formed in a similar disproportionation reaction between hydrogen atoms and alkyl radicals. Since allyl radicals normally decay via recombination (*Table 5.1*), the presence of double bonds increases

Table 5.1 **Comparison of the ratio of the rate constants for recombination (k_r) and disproportionation (k_d) of isopropyl and allyl radicals in the gas phase**

Radical	k_r/k_d
$CH_3\dot{C}H\,CH_3$	1.7
$\dot{C}H_2CH=CH_2$	125.0

the crosslinking efficiency. This mechanism may also explain the effect of acetylene in increasing the crosslinking yield in irradiated polyethylene. The addition of acetylene to secondary alkyl radicals leads to the appearance of side-chain allylic radicals, which are more mobile. The effect of the mobility of side groups in polyethylene becomes most pronounced at temperatures just below the melting point of polyethylene crystallites (*Fig. 5.4*).

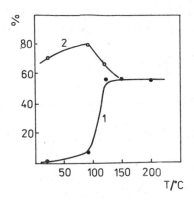

Fig. 5.4. Effect of temperature on the amount of gel formed during the irradiation of polypropylene in vacuo (*1*) and in acetylene (*2*)

(According to the results of [13])

The increase of the crosslinking efficiency at 150 °C in vacuo to a *G*-value of ca. 5 indicates crosslinking via double bonds in anomalous functional groups of the polyethylene.

From this viewpoint, those monomers having several C=C double bonds would be expected to increase the yield of crosslinking in the irradiated polymer [14]. The effect of easier contact of reactive radicals is augmented by the presence of double bonds in the polymer chain. At these sites, crosslinking is induced even by addition of terminal macroradicals formed on primary scission of a main chain.

The influence of the arrangement of the macromolecules on the formation of crosslinks in the solid state is noteworthy. The effect of the same radiation dose both on a compact bulk sample and a fine powder of perfectly crystallized polyethylene, which are chemically identical, is to induce approximately equal concentrations of secondary alkyl radicals. The effective density of crosslinks in the perfectly crystalline sample is, however, lower. To achieve the same degree of crosslinking, a ten-fold higher dose of radiation is necessary for the crystalline sample. This is also an indication that the rate of recombination of alkyl radicals in crystalline regions is lower than that in amorphous regions.

The preparation of monocrystals as carried out from dilute solutions leads to

105

a preponderant appearance of segments of the same macromolecule on the monocrystal surface (*Fig. 5.5*), while the surfaces of lamellae crystallized from the melt also involve segments of different macromolecules. As a consequence,

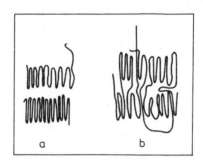

Fig. 5.5. Lamellas of crystallites prepared from a linear polymer crystallized
(a) from solution and (b) from the melt

in the decay of macroradicals in the amorphous phase of the monocrystal surface, small intramolecular rings are formed which do not decrease the solubility of the polyethylene. The different efficiency of crosslinking of polyethylene containing various defects in the macromolecular chain [15] may be explained accordingly. The difference in the structure brings about a different supermolecular arrangement of the polyethylene chains (*Table 5.2*). The lower the crystallinity of the polymer, the higher is the yield of crosslinks; this tendency is apparent even though the plot of both parameters is not a straight line. This may be due to the appearance of intramolecular rings indicated above, which represents a certain ratio of the overall number of crosslinks. If the lamellar surface formed by folds in the macromolecule is disrupted, then intramolecular rings are produced with a lower probability and more free radicals are available for intermolecular crosslinking. This is in accordance with the higher content of substituted cyclohexane rings observed in the infrared spectra of

Table 5.2 **Radiation yield G (C—C) of crosslinking for three samples of polyethylene of different structure irradiated at 35 °C**

Linear polymer of	Number of $R^1CHR^2R^3$ per 100 CH_2	Crystallinity, %	Dimensions of lamellar crystallites/ nm	G (C—C)
Ethylene	0.1	75	22.4	0.8
Ethylene with 7 mol % of propylene	3	56	16.4	1.0
Ethylene with 3 mol % of 1-butene	2	45	10.9	2.8

crosslinked linear polyethylene when compared with ethylene copolymers. In spite of much discussion [16] concerning the location of crosslinks in semicrystalline polyethylene, the problem has not been satisfactorily resolved.

β. Low-Molecular-Mass Models

Since polyethylene is a semicrystalline polymer, changes induced by radiation may involve only one of its phases. Polyethylene also contains side branches, double bonds and other anomalous units. These facts complicate the quantitative study of the crosslinking of polyethylene considerably; from this viewpoint the models involving compounds of low-molecular mass, for which experimental methods of separation and identification of reaction products are more developed, seem to offer advantages.

It follows from an investigation of crosslinking of low-molecular-mass crystalline alkanes [17], that crosslinked molecules react faster than their noncrosslinked neighbours. The ratio of trimers and higher oligomers is therefore higher, and of dimers lower, than one should expect from random formation of crosslinks between the crystalline parts of the polymer. This is also associated with a phase separation of crosslinked and branched molecules to amorphous microregions, which increases their dimensions during irradiation, even when their number remains constant. Although the primary excitation of chemical bonds occurs randomly, the crosslinking process is inhomogeneous. The transfer of excitation energy to the defect centres of a crystal probably proceeds as an intramolecular

$$-\dot{C}H-(CH_2)_n- \quad \longrightarrow \quad -(CH_2)_n-\dot{C}H-$$

or intermolecular

$$
\begin{array}{cc}
-\dot{C}H- & -CH_2- \\
& \longrightarrow \\
-CH_2- & -\dot{C}H-
\end{array}
$$

migration of hydrogen which shifts the radical sites to the amorphous phase where they may recombine.

The analysis of appropriate ESR spectra shows that about 50 % of the alkyl radicals formed from an irradiated alkane at 4 K are found in pairs while at 77 K the figure is only 2 %. Homogenization of the distribution of radicals is apparently due to relay-like migration of hydrogen atoms. On the other hand, the primary radical pair may induce the transfer reaction of a 'hot' hydrogen atom with the surrounding matrix. Should such a hot atom react with an adjacent CH_2 group on the same chain, then a double (vinylene) bond is formed.

The effects of aggregation and arrangement of the alkane $C_{23}H_{48}$ on its

107

crosslinking yield at three irradiation temperatures may be demonstrated as follows.

Temperature/K	274	315	326
State of aggregation	crystalline	crystalline	liquid
Arrangement	orthorhombic	hexagonal	amorphous
Radiation yield			
G(C—C) (crosslinks)	1.0	2.5	1.75

The hexagonal crystalline structure is rather mobile at a temperature several degrees below its melting point but insufficiently to enable the separation of radicals from radical pairs by fast diffusion into the liquid phase. In the hexagonal arrangement only rotation of the molecules along their axis occurs.

The state of aggregation of the hydrocarbon affects not only the kinetics of the process during irradiation but also the homogeneity in the distribution of the crosslinks. This may be seen in changes of the melting temperature of crystallites; if polyethylene is irradiated in the solid state, then its melting temperature changes only slightly when the sample is successively melted and cooled down. Considerably larger changes are observed when polyethylene is irradiated in the melt and then allowed to crystallize. The reasons for this difference become obvious if we bear in mind the regular arrangement of macromolecules folded in crystallites. The crosslinking of macromolecules in the amorphous phase of semicrystalline polymer cannot influence the reverse crystallization when the polymer is subsequently melted and crystallized; the situation is quite different for the random fixation of macromolecules during irradiation of the polymer melt.

Besides the formation of crosslinks and hydrogen molecules during the irradiation of polyethylene, there are also formed *trans*-vinyl bonds due to disproportionation and dehydrogenation of hot secondary alkyl radicals. The terminal vinyl groups and vinylidene double bonds, which arise from the polymer during its preparation, disappear. This is an indication that some crosslinks are also formed by an addition mechanism; it should be remembered, however, that the main contribution to crosslinking originates from the unsaturation of the terminal vinyl bonds [18].

γ. The effect of Structure of the Macromolecule

The effect of terminal unsaturation on the crosslinking efficiency of polyethylene is augmented in the crosslinking of polydienes. While natural rubber and 1,4-polybutadiene crosslink with an efficiency comparable to polyethylene, the structural units of 1,2-polybutadiene are about 100 times more effective.

A more exact comparison of crosslinking efficiency is rather difficult since the character of the network produced also depends strongly on the physical parameters of the structure of the polymer.

Provided that carbon chain of polyethylene is regularly interrupted by oxygen atoms as for example in poly(ethylene oxide), the crosslinking efficiency is higher due to a greater mobility of the chain as compared with polyethylene at the same temperature. A similar comparison can be made for polyamides, polyesters and other heterochain polymers. The substitution of hydrogen in a vinyl polymer chain decreases the crosslinking efficiency since the corresponding substituted carbon macroradicals undergo fragmentation and disproportionation to a greater extent than unsubstituted ones. In vinylidene polymers, the crosslinking is even more suppressed.

Polymers with aromatic groups crosslink with a yield about 100 times lower. The retardation or "protecting" effect of the aromatic group may be explained both by the slower radiolysis of aromatic polymers into radicals and by the rather efficient scavenging of hydrogen atoms by the aromatic group. The addition of hydrogen to the phenyl groups of polystyrene yields a cyclohexadienyl radical

$$\text{H}^{\cdot} + \text{—CH}_2\text{CH—} \longrightarrow \text{—CH}_2\text{CH—}$$

which disproportionates with an alkyl macroradical to produce a molecule of the original polymer.

As already pointed out, because of differences in the physical properties of polymers and the presence of different anomalous structural units in the chain,

Table 5.3   The yields (G) of crosslinking (C) and degradation (D) of some polymers irradiated at 20°C.

Polymer	G	
	C	D
Cis-1,4-polybutadiene	3.8	—
Polyethylene	3.0	0.5
Polydimethylsiloxane	2.7	0.5
Poly(vinyl chloride)	2.2	—
Natural rubber	1.1	0.2
Poly(methyl methacrylate)	0.6	0.2
Polypropylene	0.6	1.1
Polycaprolactam	0.5	0.6
Poly(vinyl acetate)	0.2	0.06
Polystyrene	0.05	0.02

a general quantitative comparison of the crosslinking efficiencies of mers of different chemical structure is virtually impossible. However, we offer some comparisons for some linear polymers (*Table 5.3*) incorporating the data of [19].

In addition to the chemical and physical structure of the polymer, the crosslinking process may be influenced by certain radical scavengers present in the polymer as impurities, initiator residues, oxygen, or as the result of atmospheric pollution. The retarding effect of some inhibitors on the formation of crosslinks is one of the numerous pieces of evidence [20] that *radiation crosslinking has a radical nature*. The structure of the crosslinks, which is not always identical, is also determined by the parent polymer itself. This may be demonstrated by poly(dimethyl siloxane) rubbers where three kinds of crosslinks were identified, namely

$$CH_3\overset{|}{\underset{|}{Si}}CH_2\overset{|}{\underset{|}{Si}}CH_3 \qquad CH_3\overset{|}{\underset{|}{Si}}\overset{|}{\underset{|}{Si}}CH_3 \qquad CH_3\overset{|}{\underset{|}{Si}}CH_2CH_2\overset{|}{\underset{|}{Si}}CH_3$$

$$1 \qquad : \qquad 1 \qquad : \qquad 0.6$$

The crosslinked product of recombination of unlike macroradicals is that most frequently encountered. It is formed via the reaction of primary methyl radicals with the methyl groups of adjacent chains and after the recombination of silicon-centred and methylene radicals. Dimethylene crosslinks are always accompanied by hydrogen abstraction, while the combination of two silicon radicals is preceded by the abstraction of two methyl groups from neighbouring macromolecules.

The analytical differentiation of individual types of crosslink in the networks of other polymers is more complicated and requires further detailed analysis.

2. Photochemical Crosslinking

The crosslinking reaction of macromolecules initiated by light, which occurs only in the thin surface layer, is used practically for the production of chips, electrical microcircuits and offset polygraphic plates [21, 22].

Photosensitive polymers are particularly advantageous for such a purpose. In the presence of 0.1—1 % photoinitiator or photosensitizer, even polyethylene, which itself is relatively resistant to light, may be crosslinked. Benzophenone may be used as a photoinitiator; in its excited state is able to abstract hydrogen from a hydrocarbon chain

$$(C_6H_5)_2CO^* + {-}CH_2{-} \rightarrow (C_6H_5)_2\dot{C}OH + {-}\dot{C}H{-}$$

and the recombination of the resultant macroradicals gives the required crosslink. In addition to crosslinks, *trans*-vinylene double bonds are again formed

110

and vinyl and vinylidene double bonds disappear from the system. The picture is qualitatively the same as obtained by the use of the far more energetic gamma radiation. The appearance of unsaturation in irradiated polymers seems therefore to be due to the subsequent reactions of radicals, such as disproportionation and fragmentation, and not just to the direct interaction of gamma rays and polyethylene.

Instead of ketone sensitizers, other simple compounds such as tetrachloroethylene [23] may be used. In the presence of the latter compound in polyethylene, 253.7 nm light induces the simultaneous formation of the crosslinked polyethylene, trichloroethylene and HCl. In the presence of air oxygen, phosgene and carbonyl groups on the polymer backbone are formed. Other polymers which have methylene groups in their structural units (aliphatic polyamides, polyacrylates, etc.) crosslink in the same way.

In this field, the crosslinking of poly(vinyl alcohol) esters of cinnamic acid is technically important. This photodimerization of the cinnamic acid moieties to cyclobutane rings resembles the (2 + 2) cycloaddition of substituent chalcone groups

$$2\left|-C_6H_4-CH=CHCOC_6H_5\right|\xrightarrow{h\nu}\left|\begin{array}{c}C_6H_5\\CO\\CH-CHC_6H_4-\\-C_6H_4CH-CH\\CO\\C_6H_5\end{array}\right|$$

Polymers with attached anthracene units undergo analogous (4 + 4) photocycloaddition.

Other reactive groups bound to the polymer chain which take part in photocrosslinking are azides and sulphazides

$$\left|-OCONH-\bigcirc\!\!-SO_2N_3\right| + \overset{|}{\underset{|}{C}}H_2 \xrightarrow{h\nu} \left|-OCONH-\bigcirc\!\!-SO_2NH-\overset{|}{C}H\right| + N_2$$

Also photoreactive groups such as

RO— ⟨ ⟩ C=O   RO—△—NCO

introduced into macromolecules can participate in photocrosslinking reactions [24] and may be potentially useful.

The crosslinking of water-soluble polymers such as gelatine, proteins, poly-saccharides, poly(vinyl alcohol), poly(vinyl pyrrollidone), etc. mixed with dichromate ions is empirically well-elaborated and extensively used in practice, even though its mechanism has not been established unequivocally. The simple view is that Cr(VI) ions are reduced by light to Cr(III) ions which crosslink the macromolecules via coordination to hydroxyl or carboxyl side-groups. The mechanism of the crosslinking resembles that in leather tanning where Cr complexes are reduced chemically. The drawback of the dichromate systems lies in the short shelf-life of the photosensitive blends (24 hours at maximum) since chemical crosslinking proceeds even in the dark.

3. Peroxide-induced Crosslinking

Peroxides are widely used as chemical initiators for radical crosslinking. In nonpolar media, they dissociate to alkoxyl radicals

$$R^1O\!-\!OR^2 \quad \rightarrow \quad R^1O^{\cdot} + R^2O^{\cdot}$$

which abstract hydrogen from the surrounding macromolecules. The subsequent reactions are similar to those established for radiation or photochemical crosslinking. Compared with radiation initiation, however, the macroradicals are formed here more selectively and during the crosslinking of the above-mentioned polydimethylsiloxane by dibenzoyl peroxide, dimethylene crosslink bridges are formed predominantly. The decomposition of peroxides in polymers [25] is somewhat slower when compared with low-molecular-mass media but not

Table 5.4   Rate constants for the decomposition of dibenzoyl peroxide in some polymers and in benzene at 75 °C

Polymer	Atmosphere	$k/10^{-6}\,s^{-1}$
Poly(vinyl chloride)	A	3.7
Polyethylene	A	3.4
	N_2	5.6
Polystyrene	A	5.5
	N_2	16
Polyamide 66	A	10
	He	23
Polyformaldehyde	A	12
	V	53
Benzene	N_2	17.5

A = air. V = vacuum. N_2 = nitrogen

112

as slow as it might be expected from the large difference in viscosity of the two types of media. The relatively small difference in the rate constants for spontaneous decomposition of benzoyl peroxide (BP) in different polymers (*Table 5.4*) may be interpreted in terms of the several alternatives to the escape of the primary radicals R from the solvent cage

$$BP \;\rightleftharpoons\; [R^{\cdot}\; {}^{\cdot}R] \;\xrightarrow{k_D}\; R^{\cdot} + R^{\cdot} \qquad\qquad R = C_6H_5COO^{\cdot}$$

$$\Big\downarrow k_i$$

$$R{-}r \;\longleftarrow\; [R^{\cdot} + r^{\cdot}] \;\xrightarrow{k_D}\; R^{\cdot} + r^{\cdot} \qquad\qquad r = \dot{C}_6H_5$$
$$\underset{CO_2}{}$$

The slow diffusion of radicals from their place of origin in the polymer medium decreases the extent of induced decomposition of the peroxide as compared with low-molecular solvents but even so some chain decomposition still occurs.

Induced decomposition may start with radicals derived from dibenzoyl peroxide (benzoyloxy, phenyl, diphenyl, and cyclohexyl radicals)

$$R^{\cdot} + BP \;\rightarrow\; RX + R^{\cdot}$$

or with radicals derived from the polymer medium

$$P^{\cdot} + BP \;\rightarrow\; PX + R^{\cdot}$$

The ratio of these alternatives depends on the character of the polymer. The decomposition of benzoyl peroxide induced by alkyl radicals is evident from the decrease of the rate constant of the process in the presence of oxygen, which converts alkyl radicals to peroxyl radicals which are less reactive towards peroxide molecules.

The higher rate constants for peroxide decomposition in more polar polymers parallels the same trend in low-molecular-mass solvents.

A slight decrease in the rate of decomposition was observed for dicumyl peroxide in isotactic polypropylene below the melting temperature of the polymer crystallites.

Even smaller differences between the decomposition of peroxide in a polymer and its low-molecular analogue are observed when the polymer is in a highly elastic or plastic state. The deceleration of peroxide decomposition in a polymer medium seems, therefore, to be due to a slower primary dissociation of the peroxidic bond.

α. Crosslinking Efficiency

As in the case of radiation crosslinking, the number of crosslinks formed in the polymer relative to the amount of peroxide decomposed, depends generally on

113

Table 5.5   Fraction of insoluble polyethylene obtained after thermal decomposition (15 min, 185 °C) of 2 % w. of peresters $R(CO(OOC(CH_3))_3$

R in perester	%	R in perester	%
$C_6H_5.C_3H_5$ [a]	96	C_6H_{11} [b]	0
$CH_3(CH_2)_2$	91	$(CH_3)_2CH$	0
$CH_3(CH_2)_{10}$	73	$(CH_3)_3C$	0
$CH_3(CH_2)_6—CH=CH$	68	$C_6H_5CH_2$	0

a) cyclopropyl. b) cyclohexyl

the structure of the polymer chain. Scission of the main chain is less frequent here but it cannot be excluded. Taking the familiar example of the peroxide crosslinking of polyethylene and of its low-molecular-mass models, as expected, the yield of crosslinks depends on the structure of the radical initiator. Thus small changes in the structure of a given peroxidic compound such as a perester may reduce the crosslinking efficiency to zero [26] (*Table 5.5*). (Compare for example the *tert*-butyl peresters of butyric and isobutyric acid where 91 and 0 % of gel is formed respectively at 185 °C after 15 min and at 2 mol % of perester.)

Analysis of the decomposition products enabled formulation of the following mechanism

$$R(CO)OOC(CH_3)_3 \rightleftharpoons R(CO)O^{\bullet} + {}^{\bullet}OC(CH_3)_3 \rightleftharpoons \dot{C}H_3 + CH_3COCH_3$$

$$\Updownarrow$$

$$R^{\bullet} + CO_2$$

The oxy-radicals are the most reactive participants in transfer reactions to the hydrocarbon chain. Provided that the transfer reactions of oxy-radicals compete successfully with their fragmentation, then crosslinking will be favoured. In such a case, two macroradicals will be formed in close proximity and may form a crosslink by combination although a fraction of the macroradical pairs decay via disproportionation. Some macroradicals separate by relay-like transfer reactions with surrounding macromolecules. A less favourable situation arises in the event of fragmentation of at least one primary oxy-radical of the initial pair. Since the carbon-centered radical produced is less reactive in transfer reactions than the oxy-radical, the probability of its decay with its partner is higher. This eliminates any possibility of the transfer to polymer and of the subsequent combination of two macroradicals, which is the probable reason for the low efficiency of crosslinking of polyethylene with some peroxides.

Early papers on the crosslinking of polyethylene usually commented that the number of crosslinks determined from the degree of solubility of modified polymer is approximately equal to one half of the molecules of peroxide decom-

114

posed (particularly with reference to dicumyl peroxide). From the dimerization of low-molecular-mass alkanes [27] it was established that the efficiency is not 1 but only 0.5. A similar value was also obtained for *tert*-butyl perbenzoate and pentadecane at 150 °C. About 30 % of initiator radicals are consumed in formation of double bonds, 15 % self decay and about 5 % recombine with alkyl radicals derived from the alkane. Secondary radicals of an alkane do not fragment to a measurable extent which indicates that the analogous reaction in polyethylene may occur only at anomalous structural units.

In accounting for the reasons for the difference in crosslinking efficiency between polyethylene and its low-molecular-mass models, we may assume that the terminal vinyl and vinylidene double bonds or *physically entangled macromolecules* which are equivalent to chemical crosslinks also contribute to the formation of crosslinks. The higher viscosity of the polymer is, moreover, likely to enhance the formation of macroradical pairs due to slower diffusion of the primary radicals away from their place of origin. The fragmentation reactions of primary radicals to form less reactive radicals will also be lessened in more viscous media.

The efficiency of crosslinking of technical mixtures which contain additives (fillers, lubricants, antioxidants, etc.) is usually lower. In some cases, crosslinking is enhanced by addition of polyfunctional monomers which also optimize the consumption of peroxide.

The effect of the structure of the polymer chain has already been pointed out for radiation crosslinking. The tendency of macroradicals to undergo fragmentation or disproportionation or to enter relay-like transfer-reactions affects the crosslinking efficiency considerably.

Thus the copolymerization of polypropylene with ethylene decreases the crosslinking efficiency according to the following data

mol % C_3H_6 in copolymer with ethylene	25	50	90
crosslinking efficiency, %	70	50	10

From the same reasons, the crosslinking of polypropylene requires a 20-fold greater amount of peroxide as compared with polyethylene [28] and polyisobutylene undergoes only degradation in the presence of oxy-radicals.

Ethylene-propylene elastomers, which are important from a commercial point of view, are terpolymers of ethylene (60 to 80 mol %), propylene and a diene monomer (ethylidenenorbornene or 1,4-hexadiene, 1 to 3 mol %). The thermoviscoelasticity of these elastomers depends not only on the density of chemical crosslinks. Their complex behaviour is a result of the presence of crystalline domains, in particular at a low density of crosslinks for samples with a high ethylene content [29].

The crosslinking of polydienes with peroxides has a similar mechanism and efficiency as that of polyethylene. Provided that the polydiene has vinyl side-groups, the efficiency of crosslinking increases above 1 and is directly proportional to the content of side double-bonds. However, the crosslinking of polydienes is only of small practical use since it cannot compete successfully with more developed and economical procedure of vulcanization by sulphur.

β. The Choice of Peroxide

The choice of peroxide used for the crosslinking of polymers is determined predominantly by the efficiency of the process at a given processing temperature. Equally important are other considerations having no direct connection with crosslinking, such as the safety towards handling of the peroxide, its toxicity, its decomposition temperature, etc. Dibenzoyl peroxide which is largely used for the initiation of vinyl polymerization is less suitable as a crosslinking agent because it decomposes at a relatively low temperature which raises problems with the processing of the polymer. The relatively high degree of induced decomposition of dibenzoyl peroxide makes it inefficient as regards crosslinking, discouraging its application.

The first technological procedures for the crosslinking of polyethylene and natural rubber were based upon dicumyl peroxide. This peroxide, however, produces acetophenone in its decomposition which should be avoided in technological practice because of its smell. This property is reduced when using 1,4-di-*tert*-butylperoxy-diisopropylbenzene. *Tert*-butyl perbenzoate and other dialkyl peroxides such as 2,5-dimethyl-2,5-di-*tert*-butylperoxyhexane are also suitable for crosslinking polymers. The duration of crosslinking with a peroxide

Table 5.6   **Temperature of decomposition for a given half-life $\tau/2$ of the commonest peroxides in benzene solution**

Peroxide	°C[a] $\tau/2 = 10\,h$	°C[a] $\tau/2 = 1\,h$	°C[b] $\tau/2 = 1\,min$
Tert-butyl hydroperoxide	172	195	243
2,5-Dimethyl-2,5-di(*tert*-butylperoxy)hexene-33	128	149	192
Di-*tert*-butyl peroxide	126	149	193
2,5-Dimethyl-2,5-di (*tert*-butylperoxy)hexane	119	138	178
Dicumyl peroxide	115	135	171
Tert-butyl perbenzoate	105	125	165
Dibenzoyl peroxide	73	92	132
Dilauroyl peroxide	66	84	122
2,4-Dichlorodibenzoyl peroxide	54	72	108

a) measured values. b) calculated values

is usually chosen to equal four times the half-life for the peroxide decomposition (*Table 5.6*). The decomposition of the residual 5 % is lengthy and does not contribute effectively to the degree of crosslinking. At low concentrations of peroxide, the stationary concentration of radicals in the system will be low which will favour the occurence of monomolecular side-reactions detrimental to the bimolecular recombination of the macroradicals. From this aspect, the faster decomposition of the peroxide at a higher temperature is to be recommended.

4. Crosslinking in Dilute Solutions of Polymers

The dilution of a polymer with a solvent decreases the crosslinking efficiency through side-reactions of the macroradicals or by interaction of the crosslinking agents with the solvent, and at a certain concentration of polymer (usually below 0.5 %), crosslinking fails to occur.

During radiation crosslinking of water-soluble polymers such as poly(ethylene oxide) and poly(vinyl alcohol), the apparent paradox is found of an increase of crosslinking efficiency with dilution of the polymer. The increased crosslinking efficiency, calculated from the number of crosslinks per given radiation dose, may be explained by the subsequent reactions of radicals from the medium with dissolved polymer. The efficiency of chemical crosslinking expressed by the number of crosslinks per unit amount of a radical source (peroxide), however, will always decrease with dilution. At a critical dilution, the crosslinking efficiency drops to zero. This rapid decrease of efficiency follows from a turnover in reaction mechanism when the intramolecular linking of coil segments of the macromolecules prevails over the intermolecular formation of crosslinks between adjacent macromolecules. Each intramolecular crosslink means not only the loss of one potential intermolecular linkage, but it also contributes to a greater contraction of the polymer coil and hence to a decrease in the probability of crosslinking. The small number of crosslinks may be formed only as local microgels at the beginning of the process in the case of dilute polymers solutions. The conformations of the developing crosslinked polymer are influenced by reaction conditions such as the radiation dose (or the amount of peroxide), the polymer concentration and its molecular mass, the pH, and the presence of additives.

The crosslinking of more concentrated solutions provides another interesting aspect. The gel structure appears when an average of one crosslink per macromolecule is achieved. At this stage of reaction, a small amount of polymer can absorb about a 50-fold higher mass of the solvent. Crosslinking thus changes the swelling of a polymer considerably and increases the heterogeneity of the system. From a homogeneously swollen gel, the solvent phase, which is initially uniformly distributed throughout the whole volume, is gradually excluded and the spongy macrostructure of the crosslinked polymer results [30].

When using more reactive organic solvents, the crosslinking reaction is always accompanied by combination reactions of solvent radicals with macroradicals and crosslinked macromolecules become grafted with solvent [31].

If the polymer has reactive functional groups, the bifunctional reactants may be used for crosslinking in the solution. In such a way, compact aqueous gels from soluble polysaccharides may be prepared [32].

B. INTERPENETRATING NETWORKS OF UNLIKE POLYMERS

The simplest way of modifying the properties of polymers consists of mixing polymers of different properties. In most polymer blends, however, the different macromolecules are phase-separated. High polymers are usually incompatible even in the same solvent, and on mixing two solutions of different polymers dissolved in the same solvent, two phases appear quickly, each of them containing one polymer.

Phase separation may be eliminated by crosslinking of the mixed polymer chains [33, 34]. It is sufficient if the two types of macromolecule are crosslinked themselves, because the interpenetration of the chains will prevent phase separation. Interpenetration and mixing of polymer segments may be so perfect that the product is distinguished from a copolymer only with difficulty, having for example only a single glass transition temperature. After disruption of the network of one polymer and its subsequent extraction, the network of the second polymer may be preserved with properties corresponding to when it is prepared independently.

Interpenetrating networks may be prepared in several ways. The swelling of a crosslinked polymer by a chemically different monomer which is subsequently polymerized in situ in the presence of a polyfunctional comonomer is the most frequent.

For example, poly(2,6-dimethyl-1,4-phenylene oxide) is thus brominated and subsequently crosslinked with ethylene diamine

The crosslinked polymer is then swollen with styrene containing a small amount of divinyl benzene, and the mixture of monomers is polymerized.

118

Another approach involves the mixing of a linear polymer with oligomer and selective crosslinking agents. The reaction of mutually dissolved but independently reacting monomers or oligomers giving the crosslinked polymer [35] appears to be even more advantageous. Such a procedure is exemplified by the crosslinking in a mixture of acrylates with ethylene glycol dimethacrylate as crosslinking agent and dibenzoyl peroxide as one reaction system and with oligomeric poly(1,4-oxybutylene glycol) mixed with an equivalent amount of 4,4-methylene-*bis*(cyclohexyl isocyanate) and trimethylol propane as crosslinking agent as the other reaction system. Since the molecular masses of the individual components of the system are low, the original mixture of monomers and oligomer forms a single phase. During reaction, the thermodynamic incompatibility of the two systems increases. Phase separation is diffusion-controlled and its appearance requires a certain time. Provided that the course of the growth reaction of the crosslinked polymers is fast, the necessary diffusion of the polymer chains cannot be fully realized. After complete reaction and crosslinking, the phase separation cannot proceed because it is prevented by the polymer network (*Fig. 5.6*). The faster the crosslinking of the reaction system, the more

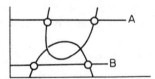

Fig. 5.6. Catenane-type entanglement of two unequal polymer chains

homogeneous will be the resulting network, with smaller domains of identical segments of macromolecules. In the less common case of miscible polymers such as poly(vinyl chloride) and butadiene-acrylonitrile copolymers, the suitable choice of a crosslinking procedure [36], such as vulcanization by sulphur is quite sufficient. In the case of compatible polymers, there is no danger of phase separation and the denser crosslinking of the elastomer component of the polydiene appears to be quite sufficient. The mixing of different polymers in molten form is energetically rather expensive, and the mixing of latexes and the simultaneous precipitation is preferable.

In general therefore the preparation of interpenetrating polymer networks requires either the preparation of different kinds of crosslinks in a stepwise process or, when appropriate, the utilization of independent reaction mechanisms.

Most studies to date deal with the crosslinking of mixtures of organic compounds but some crosslinking experiments have been applied to inorganic glasses [37].

119

C. THE EXTENSION OF MACROMOLECULES AT THEIR END GROUPS

A special way of lengthening macromolecules is by condensation or addition reactions of their end groups [38]. Such a reaction may occur if there is sufficient probability of the end groups firstly being involved in collisions and secondly, being moderately reactive. Both conditions are met in the crosslinking of two-component siloxane rubbers with identical structural mers; one part of the siloxane macromolecules has vinyl or allyl double bonds while the second part has silyl hydride end-groups. The addition reaction of these functional groups

$$
\begin{array}{ccc}
\overset{R}{\underset{R}{|}} & \overset{R}{\underset{R}{|}} & \overset{R}{\underset{R}{|}}\;\;\overset{R}{\underset{R}{|}} \\
-\text{SiCH}{=}\text{CH}_2 + \text{HSi}- & \longrightarrow & -\text{SiCH}_2\text{CH}_2\text{Si}- \\
\end{array}
$$

crosslinks the branched macromolecules. The rate of the reaction may be controlled either by temperature or catalytically.

Polyfunctional crosslinking agents may be used to crosslink linear macromolecules via their end groups. One example of such a crosslinking procedure is the reaction of end hydroxyl groups with tetra- or trialkoxy silanes and with tri-isocyanates. There are numerous similar examples of this approach, which have been studied intensively.

The reactions of end groups of macromolecules of different primary structure should produce block copolymers. Such reactions, however, do not usually occur since the probability of collision of the end groups is rather low because of the repulsive interactions of the macromolecular coils. Block copolymers are better prepared by the chain reaction of the end groups of macromolecules with low-molecular-mass monomers. Thermally unstable functional groups which give macroradicals, or polar groups initiating polymerization after heating, may be used as 'end-macroinitiators' of a chain reaction of monomers [39, 40]. The polymerization may also be initiated directly through nondeactivated (living) ionic sites which survive in the system from the preparation of the original polymer. Anionic polymerization is the most suitable for the preparation of equally reactive end groups, which give anion-radicals in the initiation stage and finally dianions on termination. Dianions initiate the polymerization of the monomer present until its complete consumption and the polymer dianionic sites remain conserved in the system. These sites may be, of course, easily deactivated by water, CO_2, alcohols and many other compounds and transformed to more stable end-groups. From the living polymer dianion B and fresh reactive monomer A, successive polymerization leads to the triblock copolymers A_mBA_n. Such a procedure based on polydiene dianions B, and with styrene as A, is used in the synthesis of thermoplastic rubbers.

120

Hydroxyl end-groups may be used as initiators for the subsequent ionic polymerization of epoxides (oxiranes). This method is applied to the preparation of the surfactant copolymers derived from propylene oxide and ethylene oxide

$$HO(CHCH_3CH_2O)_mH + 2n\ CH_2CH_2 \xrightarrow{\hspace{1cm}} HO(CH_2CH_2O)_n(CHCH_3CH_2O)_m(CH_2CH_2O)_nH$$
$$\underset{O}{\diagdown\diagup}$$

polyoxamers

$$HO(CH_2CH_2O)_sH + 2p\ CHCH_3CH_2 \xrightarrow{\hspace{1cm}} HO(CHCH_3CH_2O)_p(CH_2CH_2O)_s(CHCH_3CH_2O)_pH$$
$$\underset{O}{\diagdown\diagup}$$

meroxapols

Besides the linear copolymers of propylene oxide and ethylene oxide, branched copolymers with two block branches fixed to polyamides can be produced, for example polyamines with four block branches fixed to ethylene diamine [41].

The interaction of oligomeric dications with structurally different macromolecular dianions [42] offers new possibilities for modification of polymers. On mixing solutions of dianions of poly(dimethyl siloxane) with dications of polytetrahydrofuran, a block copolymer consisting of many repeating polymer blocks in the copolymer chain is formed; such a copolymer cannot be synthesized in any other way.

The chemical linkage of macromolecules having either identical or different primary structures either inside or at the chain-ends induces quite significant changes in properties which are related to the decrease in the concentration of structural units in the polymer system. As regards the linkage of macromolecules via a covalent bond, less than 0.1 mol % of structural units is sufficient to change the polymer from the plastic to the elastic state, to cause its insolubility, etc. Similarly small changes of chemical structure are required for the synthesis of block copolymers from different polymer reagents. Large changes in physical properties due to perturbations of some fraction of the structural mers in a macromolecule follow from the fact that these induce different arrangements and mobility of the macromolecules at the supermolecular level.

References

1. RADO, R.: The Transformation Reactions of Polymers induced by Peroxides, (in Slovak). Publishing House Alfa, Bratislava 1970.
2. BALCAR, H.: Homogeneous Metathesis of Unsaturated Hydrocarbons, (in Czech). Chem. Listy 78, 35—53, 1984.
3. DONCOV, A. A., TARASOVA, Z. N., SHERSHNEV, V. A.: Some New Concepts in Crosslinking Vulcanization of Elastomers, (in Russian). Colloid. Zh. 35, 211—225, 1973.

SCHNECKO, H.: Bedeutung und Aufbaumöglichkeiten von Netzwerken. Angew. Makromol. Chem. *76/77*, 1—23, 1979.

4. HUMMEL, K.: Nachtragliche Vernetzung von Polymeren. Angew. Makromol. Chem., *76/77*, 25—38, 1979.

DONCOV, A. A., PROYTCHEVA, A. G., NOVITSKAYA, S. P., DOGADKIN, B. A.: The Effect of ZnO Vulcanization of Saturated Polyolefines by Dibenzoylthiazyldisulphide, (in Russian). Vysok. Soed., Short Comun. *13*, 671—675, 1971.

5. KUNIAK, L., MARCHESSAULT, R. H.: Study of the Crosslinking Reaction between Epichlorhydrin and Starch, Die Starke *24*, 110—116, 1972.

KUNIAK, L.: Crosslinking of Microcrystalline Cellulose with Epichlorhydrin. Cellulose Chem. Technol., *8*, 255—262, 1974.

LUBY, P., KUNIAK, L.: Crosslinking Statistics. Relative Reactivities of Amylose Hydroxyl Groups. Macromol. Chem. *180*, 2213—2220, 1979.

6. CANTOR, CH. R., SCHIMMEL, P. R.: Biophysical Chemistry. I. The Conformation of Biological Macromolecules. W. H. Freeman and Co., San Francisco 1980, p. 70.

RENCOVÁ, J.: The Crosslinks in Collagen, (in Czech). Chem. Listy, *75*, 1185—1201, 1981.

BURCHARD, W.: Networks in Nature. Brit. Polym. J., *17*, 154—163, 1985.

7. SCHULZ, G. E., SCHIRMER, R. H.: Principles of Protein Structure, Springer Verlag, New York 1979, p. 62.

HORÁKOVÁ, M., DEYL, Z.: Elastin, (in Czech). Chem. Listy, *77*, 277—287, 1983.

8. WALTON, A. G., BLACKWELL, J.: Biopolymers. Academic Press, New York 1973, p. 437.

9. BLAŽEJ, A., DEYL, Z., ADAM, M., MICHLÍK, I., SMEJKAL, P.: Structure and Properties of Fibrous Proteins, (in Slovak). Publishing House Veda, Bratislava 1978.

10. DUŠEK, K.: Formation and Structure of End-Linked Elastomer Networks. Rubber Chem. Technol., *55*, 1—22, 1982.

11. PHILIPP, B., DAWYDOFF, W., LINOW, K. J.: Polyelektrolytkomplexe-Bildungweise, Structur und Anwendungsmöglichkeiten. Z. Chem., *22*, 1—13, 1982.

12. BAIRD, W. G. Jr., JOONASE, P., ROSE, A. B., HELMAN, W. Ph.: Bibliographies on Radiation Crosslinking of Polymers. Radiat. Phys. Chem., *19*, 339—360, 1982.

DOLE, M.: History of the Irradiation Cross-Linking of Polyethylene. J. Macromol. Sci., A *15*, 1403—1409, 1981.

ANDREOPOULOS, A. G., KAMPOURIS, E. M.: Mechanical Properties of Crosslinked Polyethylene, J. Appl. Polym. Sci., *31*, 1061—1068, 1986.

13. MITSUI, H., HOSOI, F., KAGIYA, Ts.: Accelerating Effect of Acetylene on the Gamma-Radiation-Induced Cross-Linking of Polyethylene. Polymer J., *6*, 20—26, 1974.

14. LEE, D W., BRAUN, D.: Strahlenvernetzung von Polyethylen in Gegenwart von Polymerisierbaren Monomeren. Angew. Makromol. Chem., *68*, 199—211, 1978.

15. DZHIBGASHVILI, G. G., SLOVOCHOTOVA, N. A., LESCHCHENKO, S. S., KARPOV, V. L.: The Investigation of the Effect of Chemical and Supermolecular Structure on Radiation Chemical Processes in Some Polyolefines, (in Russian). Vysok. Soed., 1087—1096, 1971.

NISHIMOTO, S., KAGIYA, V. T.: Flexible Methylene Chain Length in Crosslinked Networks as a Measure of the Brittleness of Gamma Irradiated Low Density Polyethylene Sheet, Polym. Degr. Stab., *15*, 237—249, 1986.

16. HOSEMANN, R., CACKOVIC, H., LOBODA-CACKOVIC, J.: Radiation induced Cross-Links in Polyethylene and their Degradation by Ozone. Makromol. Chem., *176*, 3065—3077, 1975.

PATEL, G. N., KELLER, H. H.: Crystallinity and the Effect of Ionizing Radiation in Polyethylene. J. Polym. Sci. *13*, 303—367, 1955.

17. UNGAR, G.: Radiation Effect in Polyethylene and *n*-Alkanes. J. Materials Sci., *16*, 2635—2656, 1981.

18. SILVERMAN, J., ZOEPFL, F. J., RANDAL, J. C., MARKOVIČ, V.: The Mechanism of Radiation induced Linking Phenomenon in Polyethylene. Radiat. Phys. Chem., 22, 583—585, 1983.
19. CHAPIRO, A.: Radiation Chemistry of Polymeric Systems. Interscience, New York 1962.
20. CHARLESBY, A.: Crosslinking and Degradation of Polymers. Radiat. Phys. Chem., 18, 59—66, 1981.
21. FINTER, J., LOHSE, F., ZWEIFEL, H.: Photochemical Imaging Processes for the Formation of Metallic Patterns on Polymers, J. Photochem., 28, 175—185, 1989.
22. VOLLMANN, H.: Polymere in der Drucktechnik, Angew. Makromol. Chem., 145/146, 411—440, 1986.
23. PUKSHANSKII, M. D., ZYUZINA, L. I., KHAYKIN, S. I., SIROTA, A. G., KACHAN, A. A., GOLDEN-BERG, A. L.: The Peculiarities of Photochemical Crosslinking of Polyethylene in the Presence of Tetrachloroethylene, (in Russian). Vysok. Soed. A 14, 2096—2101, 1972.
24. GREEN, G. E., STARK, B. P., ZAHIR, S. A.: Photocrosslinkable Resin Systems. J. Macromol. Sci., Chem. C. 21, 187, 1981—82.
25. LAZÁR, M.: Solid-State Reactions of Peroxides, in: The Chemistry of Functional Groups, Peroxides. Editor Patai, S., J. Wiley and Sons. Chicester 1983, p. 777—806.
26. VAN DINE, G. W., SHAW, R. G.: Crosslinking of Polyethylene by Peresters. Polymer Preprints, 12, 713, 1971.
27. VAN DRUMPT, J. D., OSTERWIJK, H. H. J.: Kinetics and Mechanism of the Thermal Reaction between tert-Butyl Perbenzoate and n-Alkanes: A Model System for the Crosslinking of Polyethylene. J. Polym. Sci., Polym. Chem. Ed., 14, 1495—1511, 1976.
28. CHODÁK, I., LAZÁR, M.: Effect of the Type of Radical Initiator on Crosslinking of Polypropylene. Angew. Makromol. Chem., 106, 153—160, 1982.
29. SCHOLTENS, B. J. R.: Thermoviscoelastic Behaviour of EPDM Elastomers in Chemical and Physical Networks, Brit. Polym. J., 17, 134—139, 1985.
30. CHARLESBY, A.: Crosslinking of Polymers by Radiation, Chimia, 22, 153—156, 1968.
31. de BOER, A., PENNING, A. J.: Polyethylene Networks Crosslinked in Solution: Preparation, Elastic Behaviour and Oriented Crystallization. I. Crosslinking in Solution. J. Polym. Sci., Polym. Phys. Ed., 14, 187—210, 1976.
32. MAHER, G. G.: Crosslinking of Starch Xanthate IV. Epoxy Resin and Diepoxide Thickeners for Xanthate and Starch. Starke, 29, 335—339, 1977.
33. FRISCH, H. L.: Interpenetrating Polymer Networks, Brit. Polym. J., 17, 149—153, 1985.
34. LIPATOV, Yu., SERGEYEVA, L.: Interpenetrating Polymer Networks, (in Russian), Publishing House Naukova Dumka, Kiev, 1979.
35. FRISCH, H. L., FRISCH, K. C., KLEMPNER, D.: Advances in Interpenetrating Polymer Networks. Pure Appl. Chem., 53, 1557—1566, 1981.
    HOURSTON, D. J., ZIA, Y.: Semi and Fully Interpenetrating Polymer Networks Based on Polyurethane-Polyacrylate Systems. J. Appl. Polym. Sci., 28, 3745—3758, 1983.
36. KŮTA, A., DUCHÁČEK, V.: Polymer Mixtures of Nitrile Rubber-Poly(vinyl chloride) and Ways of their Crosslinking (in Czech). Chem. Listy, 75, 1170—1184, 1981.
37. WARD, T. L. GOYNES, W. R. Jr., BENERITO, R. R.: Interpenetration of Silica in a Network of Cellulose and Divalent Lead to Form Glassy Polymers, in: Polymer Alloys II. (Editors) Klempner, D., Frisch, K. C. Plenum. Publ. Corp., (1980), p. 153—166.
38. VALECKII, P. M., STOROSHUK, I. P.: Block-Copolymers of Polycondensation Type, Russ. Chem. Rev., 48, 1979. Mc Grath, J. E.: Block and Graft Copolymers. J. Chem. Educ. 58, 914—921, 1981.
39. OTSU, T., YOSHIDA, M., KURIYAMA, A.: Living Radical Polymerization in Homogeneous Solution by using Organic Sulphides as Photoinitiators, Polym. Bull., 7, 45—50, 1982.

123

40. CRIVELLO, J. V., LEE, J. L., CONION, D. A.: Cyclic Silyl Pinacol Ethers. A new Class of Multifunctional Free Radical Initiators, Polym., *16*, 95—102, 1986.
41. SCHMOLKA, I. R.: A Review of Block Polymer Surfactants. J. Amer. Oil Chem. Soc., *54*, 110 —116, 1977.
42. KUČERA, M.: Macroions and their Reactions Leading to Block Copolymers, (in Czech). Chem. Listy, *77*, 1083—1105, 1983.
    KUČERA, M., BOŽEK, F., MAJEROVÁ, K., KAHLE, L.: Combination of Polymerions. Formation of Copolymer Containing Polytetrahydrofuran and Polydimethylsiloxane Blocks. Polymer. 24, 217—222, 1983.

VI. EXCHANGE REACTIONS OF POLYMER CHAINS

The reactions of groups on the backbone of hydrocarbon polymers are less frequent than reactions of low-molecular compounds with the side-groups of a macromolecular chain. The exchange reactions occurring with heteroatom polymers may be included in this category [1].

Consider the macromolecules AMZ and ANZ, where the polymer backbone involves the unlike groups of atoms M and N, which react according to the scheme

$$AMZ + ANZ \rightleftharpoons AOA + ZPZ$$

A and Z are end-groups; for the atoms of the macromolecular backbone the following simple relation is valid

$$M + N = O + P$$

Since neither the overall number of bonds in the main chains nor the number of end-group changes, the arithmetic mean value of the molecular masses of the reactants remains constant throughout the reaction. The length of the respective macromolecules will, of course, vary as well as the polydispersity of the reaction system. Provided that a polymer with a broader distribution curve of molecular masses than corresponding to the most probable one undergoes the exchange reaction, then its dispersity becomes narrower and more uniform; the reverse is also true; i.e. for a monodisperse polymer or polymer with a narrower polydispersity than corresponding to the most probable one, the polydispersity increases [2].

More distinct changes may be observed during exchange reactions of chemically different macromolecules. When two different polyamides are heated to 230 °C, the melting temperature of their mixture gradually settles to a value corresponding to a copolyamide (*Fig. 6.1*).

The occurrence of an exchange reaction in a mixture of different polyesters at elevated temperature is accompanied by changes in its physical properties [3].

125

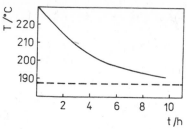

Fig. 6.1. Change in the melting temperature of a mixture of polyamide 66, the polycondensate of hexamethylene diamine and adipic acid and polycaprolactam (full line) with the time of heating at 230 °C. The dashed line denotes the behaviour of the statistical copolycondensate corresponding to the initial composition of the mixture

The mixture of polymers is firstly transformed to a block copolymer. In the final stage of the reaction, the structural units of the original polymers regularly alternate in the polymer chain. The length of their sequence depends strongly on the reaction time, the presence of catalysts and the viscosity of the reaction medium.

The chemical transformation taking place during the interaction of ester groups is illustrated by the scheme [4]

$$
R^1COR^2 + R^3COR^4 \rightleftharpoons \left[\begin{matrix} R^1C-OR^2 \\ R^3O-CR^4 \end{matrix} \right] \rightleftharpoons R^1COR^3 + R^2COR^4
$$

which is formally equivalent to the metathesis of alkanes.

Interchain exchange reactions are not specific to macromolecular compounds; thus when a mixture of ethyl stearate and cetyl acetate is heated to 120 °C, the more volatile ethyl acetate is liberated from the reaction mixture. This adversely affects the reverse esterification and cetyl stearate accumulates in the batch.

Interchain reactions of polyesters need not involve solely ester groups, but also reactions between ester groups and hydroxyl or carboxyl groups, etc.

$$
R^1COR^2 + R^3OH \rightleftharpoons \left[\begin{matrix} R^1C...OR^3 \\ R^2O...H \end{matrix} \right] \rightleftharpoons R^1COR^3 + R^2OH
$$

$$
R^1COOR^2 + R^3COOH \rightleftharpoons \left[\begin{matrix} R^1C-OR^2 \\ HO-CR^3 \end{matrix} \right] \rightleftharpoons R^3COR^2 + R^1COOH
$$

The alcoholysis of polyesters by monofunctional alcohols contributes to their degradation. An analogous mechanism operates in the regeneration of dimethyl terephthalate from poly(ethylene terephthalate) by methanol at elevated temperatures and pressures. The exchange reactions of polyesters are accelerated by p-toluenesulphonic acid or by NaOH.

Exchange reactions may also be effected between different types of heterochain polymers such as polyesters and polyamides, a process used practically in the curing of enamels and paints. If an exchange reaction of the ester groups both at the end and within the main chain occurs within a single macromolecule, then cyclic oligomers are produced. The reverse process was observed during the synthesis of polysiloxanes where cyclic monomers were transformed into linear macromolecules in the presence of catalysts.

Exchange reactions are not limited to the interaction of unlike atoms at the sites of reaction but can extend to identical atoms, e.g., various reactions of poly(alkyl sulphides). Intermolecular reactions occurring at sulphur atoms bring about a relatively fast decrease of the intrinsic strain of deformed poly-(alkyl sulphide) rubbers. Due to the exchange reaction, the macromolecular segments adopt a thermodynamically more favourable configuration as compared with the deformed state (*Fig. 6.2*). The fission of strained bonds in poly(alkyl sulphide) rubbers may also be induced by ultraviolet light.

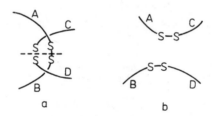

Fig. 6.2. Schematic illustration of entangled chains (*a*) of poly(alkyl sulphide) rubbers and (*b*) of the new arrangement due to exchange reactions at the polysulphide bonds

The intermolecular exchange reaction of saturated hydrocarbon polymers is less easily accomplished. To bring it about, the C—C bond must initially be cleaved by the effect of mechanical strain, ionization radiation, or by other sources of free radicals which can initiate the fragmentation of the macromolecular chains. The macroradicals formed may subsequently combine. If radicals derived from different macromolecules do so, the product is not readily distinguished from that of the exchange reaction.

In spite of the relatively easy experimentation involved, the interchain reactions of heterochain macromolecules have been studied infrequently. However, they have found practical application in the curing of some multi-component paints. Of general interest is the fact that a considerable change in the macromolecules involved may occur even at the first reaction step. It should be remembered that the final products of the modification reaction, which have an equilibrium distribution of structural units in the polymer chain, may also be obtained by direct synthesis from the monomer mixture.

These considerations concerning the possible applications of exchange reactions in the fields of paints, cements, rubbers or thermoplasts might end this chapter. However, the picture would be incomplete without some illustration of the impact of exchange reactions on the properties of biomacromolecules. It should be stressed how significantly the functional properties of, say, deoxyribonucleic acid (DNA), which transmits genetic information through a particular sequence of its four constituent bases, may be affected by such an exchange. A change in the genetic makeup of an organism via its chromosomes may occur in several ways [5]. In the exchange interaction of two different DNA macromolecules, a process denoted translocation or recombination, may occur where

$$\frac{A^1 \quad A^2}{B^1 \quad B^2} \longrightarrow \begin{matrix} A^1 & B^2 \\ \diagdown\hspace{-0.5em}\diagup \\ B^1 & A^2 \end{matrix} \longrightarrow \frac{A^1 \quad B^2}{B^1 \quad A^2}$$

part A of one macromolecule links altogether with part B of the second macromolecule and vice versa. An intramolecular cyclization reaction deletes satellite (cyclic) DNA

$$\underline{\quad\bigcirc\quad} \quad \overset{\text{deletion}}{\underset{\text{insertion}}{\rightleftharpoons}} \quad \underline{\qquad\qquad} \quad \bigcirc$$

which becomes detached from the DNA moiety and the information content of DNA is thus lowered. The reverse process of insertion incorporates a new nucleotide segment into DNA. Inversion is a similar reaction

$$\underset{n}{\underline{\quad A \qquad B \quad}}\bigcirc\overset{1}{\underset{2}{}} \longrightarrow \underset{n \ldots 4\ 3\ 2\ 1}{\underline{\quad A \qquad\qquad B \quad}}$$

which does not eliminate cyclic DNA (plasmid) but at the sites of mutual contact of the chain within one macromolecule a rearrangement occurs similar to the translocation of two different macromolecules.

The mechanism of the exchange reactions of DNA, which resembles that of synthetic polyesters, probably involves the transesterification of phosphodiesters

$$R^1OPOR^2 + R^3OPOR^4 \rightleftharpoons \left[\begin{array}{c} R^1O_3P\!-\!OR^2 \\ R^3O\!-\!PO_3R^4 \end{array} \right] \rightleftharpoons R^1OPOR^3 + R^2OPOR^4$$

The process of exchange of genetic material in chromosomes cannot therefore, be connected with DNA fracture and with the reverse linkage of resultant fragment with another fragment [6], but it may be represented simply by an exchange transesterification reaction.

The occurrence of an interchain exchange reaction is infrequent in living organisms but may be real. It does not necessarily include only the 'illegal' recombinations but also some spontaneous mutations which do not require the enzymes of a particular cell. Transesterifications may be catalyzed by transferases.

In addition to the exchange reactions at groups of atoms within the DNA chain there are also the reactions between terminal and intrinsic chain groups to be accounted for. Such reactions may partially explain one concept of molecular genetics, namely the attachment of fragments of extraneous DNA to the gene of the chromosome by "cohesion ends". An important role will be played here by the physical interactions of complementary bases which locate the transesterification reaction at a particular site of DNA.

All the above reactions alter the sequence in DNA macromolecules and increase or decrease their overall length. Such interventions in the structure of DNA will lead to mixing and combination of some features of the parents of the next generation and may be the intrinsic reason for some mutations in living organisms.

References

1. DAVIES, T., BOSTICK, E. E., BERENBAUM, M. B.: Interchange Reactions. In: Chemical Reactions of Polymers. (Editor) Fettes, E. M., Interscience, New York 1964, p. 501—550.
2. KOTLIAR, A. M.: Effect of Interchange Reactions on Nonequilibrium Distributions of Condensation Polymers and Their Associated Molecular Weight Averages. J. Polym. Sci., Polym. Chem. Ed., *11*, 1157—1165, 1973.
3. YAMADERA, R., MURANO, M.: The Determination of Randomness in Copolyesters by High Resolution Nuclear Magnetic Resonance. J. Polymer. Sci. A—1, *5*, 2259—2268, 1967.
   KIMURA, M., PORTER, R. S.: Blends of Poly (Butylene Terephthalate) and Polyarylate before and after Transesterification J. Polym. Sci., Polym. Phys. Ed., *21*, 367—378, 1983.
4. RAMJIT, H. G.: The Influence of Stereochemical Structure on the Kinetics and Mechanism of

Ester-Ester Exchange Reactions by Mass Spectrometry. II. J. Macromol. Sci., Chem., A 20, 659 —673, 1983.

5. PAČES, V., KRČMÉRY, V., ANTAL, M.: Molecular Genetics (in Slovak), Publishing House Alfa, Bratislava, 1983.

6. WATSON, J. D.: Molecular Biology of the Gene. (Translation into Czech.). (Editor), Benjamin, W. A., Publishing House Academia Prague, 1976.

VII. THE DEGRADATION OF MACROMOLECULES

Hitherto, the formation of a reactive site, either as an ion or radical, or an electronically or vibrationally excited state, on a macromolecule has been considered to be the initial stage of further growth reactions such as crosslinking, branching, grafting and transformation of functional groups, all of which conserve the integrity of the parent polymer chain. Each reaction centre, however, may simultaneously facilitate the cleavage of the corresponding α and particularly β-bond, and bring about a decrease in the molecular mass of the polymer. A fragmentation reaction, which is the process opposite to macromolecular synthesis, has a unimolecular character and usually leads to a deterioration in the properties of the polymer.

Besides the chemical transformations of polymers associated with their fragmentation, the term destruction may be extended to the supermolecular level, thereby including physical changes leading to the disruption of a more organized structure of the polymer, which results in a worsening of its functional properties [1]. In biopolymer chemistry such processes are called *denaturation* or coagulation. When referring to the slow recrystallization of polymers, the term *"destruction"* is inappropriate even though the materials properties deteriorate; such changes in the polymer are alternatively called ageing. Only those reactions which reduce the number of chemical bonds comprising the macromolecular backbone, and hence its length are included within the category of processes leading to destruction dealt with in this chapter, and to clarify this point we shall use the term *"degradation"*.

As for other reactions of macromolecules, particularly the production of initiating species, we differentiate between photo- and photo-oxidative degradation, thermo-and thermo-oxidative degradation, radiolytic and mechanochemical degradation, etc. Polymers may also degrade under the initiating effect of ozone, peroxides, halogens and other aggressive compounds, under the effects of electric fields, plasmas and corona discharges, ultrasound, laser radiation, etc. Of special interest is high temperature degradation occuring during carbonization, i.e. with the maximum degradation of the polymer.

Degradation reactions are mostly of a free radical chain character and involve oxygen. Free radicals $\overset{.}{P}$ appearing on the main polymer chain PH due to some transformation of the primary reaction centre, are converted into hydroperoxides, the decomposition of which governs the overall course of reaction

$$P^{.} + O_2 \quad \rightarrow \quad PO_2^{.}$$

$$PO_2^{.} + PH \quad \rightarrow \quad POOH + P^{.}$$

$$POOH \quad \rightarrow \quad PO^{.} + {}^{.}OH$$

The decrease in molecular mass is caused by the fragmentation of alkoxyl or, when oxygen is deficient, alkyl radicals. Before the radical disappears in a termination step, the propagation cycle may be repeated many times and a relatively small perturbance to the macromolecular system can thus have an immense impact on the properties of a polymer. In contrast to the oxidative degradation of polymers, ionic degradations are mostly of non-chain character. The initiating effects of water, acids and alkaline compounds on the cleavage of ester, ether and amido groups in heteroatom polymers occur as a sequence of equilibrium reactions.

Although the individual degradation processes are usually treated separately, many of them are closely interrelated and occur simultaneously during treatment of the polymer. During some biodegradations, the effect of the microorganisms is accompanied by initiations either with reactive compounds or with ultraviolet light; the latter promotes biodegradation as such. The long-term use of synthetic polymers is affected particularly by degradation involving light and heat with the cooperative action of oxygen, i.e. by photo-oxidation and thermo-oxidation.

A. PHOTOLYTIC AND PHOTO-OXIDATIVE DEGRADATION OF POLYMERS

The absorption of a quantum of light by a molecule occurs after a specific interaction of some functional group (chromophore) with the photon. Since bonds such as C—C, C—H, O—H and C—Cl absorb light of wavelengths shorter than 200 nm, polymers composed of these bonds should be photolytically stable and be decomposed only by shorter-wavelength UV radiation. In practice, however, virtually all polymers are sensitive to radiation of wavelengths longer than 200 nm, including those of solar radiation ($\lambda \geq 290$ nm). This is due to the carbonyl groups and alkene double bonds incorporated into the polymer by undesirable oxidation processes during synthesis, processing and storage of the polymer. These functional groups have absorption maxima between

200—400 nm. Photolysis of the polymer is, moreover, enhanced by additives and impurities which also absorb light more effectively than the polymer alone. A not insignificant contribution is made by the mechanochemical degradation of solid polymers brought about by noncompensated strains, thus the sudden extension of a non-stabilized polypropylene fibre leads to an increase in the concentration of hydroperoxides by one order of magnitude at room temperature, the resulting level being 10^{-3} mol kg^{-1}. The effects of individual chromophores (*Table 7.1*) depends on the amount of energy absorbed. It is determined by the concentration of a given chromophore, by its absorption spectrum and by the yield and reactivity of the radicals formed. Since the concentration and nature of the chromophores may change with time, the photo-reactivity of a polymer is specified by a rather broad range of quantum yields (The quantum yield is the number of chemical acts measured per photon absorbed) (*Table 7.2*) [2].

Table 7.1   **Origins of the photolability of saturated hydrocarbon polymers**

Absorbing species or chromophore for radiation of 290 nm	Origins of its presence in the polymer
C=O	Copolymerization with CO, oxidation during processing and storage
—O—O—	Oxidation during processing, reaction on storage with O_3 and $^1O_2$
—C=C—C— $\overset{\|}{O}$	Oxidation, processing
(—C=C—)n	Monomer impurity, processing
Polynuclear aromatic compounds	From surrounding atmosphere
Phenols, peroxycyclohexadienones	Phenolic antioxidants and their photo-oxidation products
C-T complex[a] polymer —O_2	Intrinsic reaction of hydrocarbon
Ions of transition metals such as Ti, Fe	Polymerization catalysts, polymer processing
TiO_2	Pigment additive

a) Charge-transfer complex

The quantum yield for bond cleavage in a polymer also depends on the segmental mobility of the macromolecule, chromophores on side-groups being more photoreactive than those on the main chain. Rather large changes in quantum yields may be observed at the glass transition temperature, T_g. For

Polymer	∅
Poly(methyl isopropyl ketone)	2.2×10^{-1}
Poly(methyl acrylate)	1.3×10^{-2}
Poly(methyl vinyl ketone)	$2 \times 10^{-2} - 2.5 \times 10^{-2}$
Poly(methyl methacrylate)	$1.7 \times 10^{-2} - 3 \times 10^{-2}$
Cellulose nitrate	$1 \times 10^{-2} - 2 \times 10^{-2}$
Poly(α-methylstyrene)	1×10^{-3}
Cellulose	$0.7 \times 10^{-3} - 1 \times 10^{-3}$
Polyamide-6	5×10^{-3}
Poly(vinylpyrrolidone)	4.3×10^{-4}
Natural rubber	4×10^{-4}
Cellulose acetate	2×10^{-4}
Polystyrene	9×10^{-5}

copolymers of methyl methacrylate and methyl vinyl ketone, the quantum yield of main chain scission increases slightly with temperature in the low temperature region — from 0.04 to 0.08; at T_g (100 °C) it jumps to 0.28 and above T_g, it attains the value in solution (0.30).

The effect of light on a chromophore M may be demonstrated by the following scheme (the superscripts 1, 2 and 3 denote singlet, doublet and triplet states of the chromophore which are characterized by 2 electrons with antiparallel spin, 1 electron and 2 electrons of parallel spin, respectively; the asterisk denotes an electronically excited state)

$$^1M + h\nu \rightarrow {}^1M^*$$
$$^1M^* + h\nu \rightarrow {}^1M^{**} \qquad \text{absorption of photon}$$
$$^1M + h\nu \rightarrow {}^2M^* + e^- \qquad \text{photoionization}$$
$$\left. \begin{array}{l} ^1M^* \rightarrow {}^1M + \text{energy} \\ ^1M^* \rightarrow {}^3M^* + \text{energy} \\ ^1M^{**} \rightarrow {}^1M^* + \text{energy} \end{array} \right\} \quad \text{nonradiative conversion}$$
$$\begin{array}{ll} ^1M^* \rightarrow {}^1M + h\nu & \quad\quad\quad\quad\quad\quad\quad \text{fluorescence} \\ ^3M^* \rightarrow {}^1M + h\nu & \text{radiative conversion} \quad \text{phosphorescence} \end{array}$$

On absorbing a photon, the ground singlet state is either converted to a mono- or doubly-excited singlet state or undergoes photoionization. An excited singlet state may give an excited triplet state in a radiationless conversion (singlet-triplet crossing) or a ground singlet state. The conversion to a ground singlet state may alternatively proceed with the emission of light as fluorescence or phosphorescence. The distribution of the population of excited states on a polymer chain is statistical. Besides the rate of formation of excited states, the production of energy from photosensitizers (F) and its loss by transfer to quenchers Q are also significant

134

$$M + F^* \rightarrow M^* + F$$

$$M^* + Q \rightarrow M + Q^*$$

These two processes prolong or shorten, respectively the lifetime of the excited molecules in the system and thus enhance or suppress the effect of photolysis.

Chemical transformations usually occur via molecular triplet states, which have relatively long lifetimes. Thus the triplet excited carbonyl group decomposes by α-scission into two radicals

$$-CH_2CH_2\overset{\cdot}{C}CH_2CH_2CH_2- \longrightarrow -CH_2CH_2\overset{\|}{\underset{O}{C}}{}^{\cdot} + {}^{\cdot}CH_2CH_2CH_2-$$

Norrish Type I mechanism

or by β-scission to methyl alkyl ketone and alkene

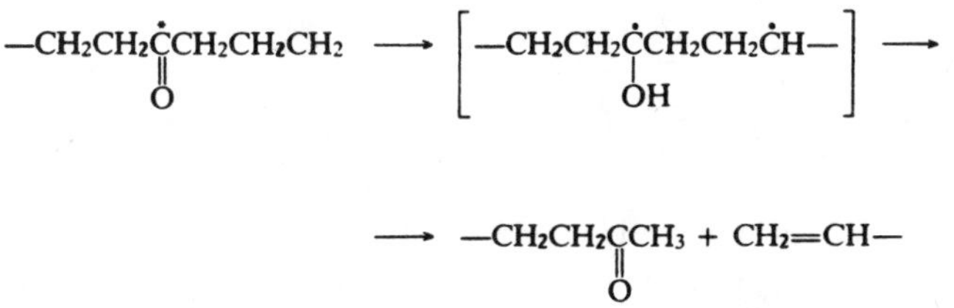

$$-CH_2CH_2\overset{\cdot}{C}CH_2CH_2CH_2 \longrightarrow \left[-CH_2CH_2\overset{\cdot}{\underset{OH}{C}}CH_2CH_2\overset{\cdot}{C}H- \right] \longrightarrow$$

$$\longrightarrow -CH_2CH_2\overset{\|}{\underset{O}{C}}CH_3 + CH_2{=}CH-$$

Norrish Type II mechanism

In the presence of oxygen, alkyl or carbonyl radicals produce hydroperoxides (POOH) or peroxyacids (PCO_3H) respectively, which on irradiation at a wavelength 300 nm decompose either directly

$$POOH^* \rightarrow PO^{\cdot} + {}^{\cdot}OH$$

or following energy transfer from excited carbonyl groups M^*

$$M^* + POOH \rightarrow M + POOH^*$$

Such energy transfer in the photo-oxidation of alkanes is important only at hydroperoxide concentrations higher than 10^{-2} mol kg^{-1}. In polymers, where —OOH and CO groups are formed in microdomains following primary oxidation, the higher local concentrations may promote such a process.

Polystyrene, which is less sensitive to photolytic degradation, is likely to decompose via hydroperoxyl groups

$$\text{(structure: } R\text{—}C(\text{O—O}\cdots\text{H—O—H})\text{—}CH_2\text{—}C\text{—}R) \xrightarrow{hv} -\overset{\displaystyle O}{\overset{\|}{C}}- + H_2O + CH_2=\overset{\displaystyle}{\underset{R}{C}}-$$

R

(R = phenyl)

The mechanism is again of a radical nature and starts with the homolysis of an O—O bond, continues by abstraction of a hydrogen atom from the methine group by the caged hydroxyl radical and ends by fragmentation of the biradical formed. The high-molecular-mass products are identical with those formed via the Norrish Type II mechanism for photolysis of the carbonyl groups of polystyrene.

The photosensitized degradation of polymers starts with the excitation of additives which subsequently react with substrate. When irradiating benzophenone in polyacetaldehyde, the oxygen atom of the excited carbonyl group of benzophenone abstracts hydrogen from the surrounding polymer

$$(C_6H_5)_2CO^{\bullet} + H\overset{\displaystyle O}{\underset{}{C}}CH_3 \longrightarrow (C_6H_5)_2\overset{\bullet}{C}OH + {}^{\bullet}\overset{\displaystyle O}{\underset{}{C}}CH_3$$

The polymer radical either fragments or reacts with oxygen and starts the chain oxidation.

Somewhat different behaviour is displayed by poly(methyl methacrylate) doped with $FeCl_3$, where $FeCl_3$ liberates a chlorine atom on irradiation

$$FeCl_3 + hv \rightarrow FeCl_2 + Cl^{\bullet}$$

The chlorine atom attacks the polymer chain and thus accelerates the photodegradation. The photodegradative effect of $FeCl_3$ is pronounced in those polymers the macroradicals of which decay via disproportionation.

TiO_2 (anatase) which is added to some polymers, especially polyamides, as a pigment has a photodegradative effect. Polycyclic aromatic hydrocarbons such as naphthalene, phenanthrene and hexahydropyrene also accelerate the photodegradation of polypropylene, but the process is stepwise. The aromatic hydrocarbons initially absorb two photons gaining access to an excited state which subsequently transfers its energy to the ground state of the carbonyl groups in the polymer. Indirectly excited carbonyl groups in the polymer chain undergo subsequent α- or β-fragmentation.

The kinetics of photo-oxidation of unstabilized polymers differs from those of thermal oxidation in the absence of an induction period (*Fig. 7.1*) for the process. In photo-oxidation, the maximum rate of reaction is usually attained

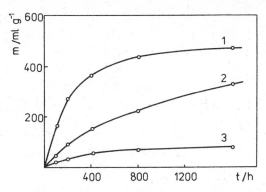

Fig. 7.1. Reaction profile of the photo-oxidation ($\lambda > 280$ nm) of (*1*) poly(4-methylpentene),(*2*) isotactic polypropylene and (*3*) polyethylene at 50 °C; *m* denotes the amount of absorbed oxygen

at the beginning of the process and then it gradually decreases. This is due to the relatively high rate of initiation. Provided that the rate of initiation is low, the autoacceleration of oxidation may be discerned. In the photo-oxidation of polystyrene initiated by shorter wavelength radiation, the induction period is not observed, while in the case of initiation at longer wavelengths, the induction period [3] (*Fig. 7.2*) is distinct. Although the photolysis and photo-oxidation of

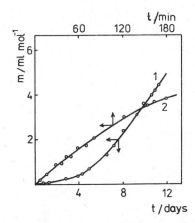

Fig. 7.2. Plot of the amount of oxygen absorbed with time in the photo-oxidation of polystyrene irradiated by ultraviolet light of wavelength (*1*) 365 nm and (*2*) 253 nm

each polymer is in some way specific, there are some general rules valid for all macromolecular compounds. It is obvious that light of shorter wavelengths will enhance the homolysis of C—C, C—H and O—H bonds, processes which will operate in parallel with those of the common chromophores (*Table 7.1*), thus the rate of oxidation will be higher as well as the amount of volatile products formed via eliminations involving the side groups. The effect of oxygen on the propagation stage of degradation, which is formally identical with the mechanism of thermal oxidation, has been noted above; in the initiation phase, oxygen may accelerate destruction by the formation of charge-transfer (CT) complexes with aliphatic chains or with traces of alkene linkages which shift the absorption maxima to longer wavelengths and decrease the photo-stability of a polymer.

Polymers such as polydienes having double bonds within their structural units will react with singlet oxygen, which is by about $100\ kJ\ mol^{-1}$ higher in energy (the state $^1\Delta_g$), than the ground state ($^3\Sigma_g^-$). Singlet oxygen transforms vinyl groups to alkylene hydroperoxides

$$^1O_2 + CH_2 = CHCH_2CH_2- \quad \rightarrow \quad HOOCH_2CH = CHCH_2-$$

which decompose more readily than alkyl hydroperoxides under irradiation at longer wavelengths.

In polymers, singlet oxygen usually originates either via energy transfer of a suitable electronically excited sensitizer to oxygen or from the disproportionation of secondary peroxyl radicals

$$2\ R^1R^2CHOO^\cdot \quad \rightarrow \quad ^1O_2 + R^1COR^2 + R^1R^2CHOH$$

Not only random carbonyl groups, but also other chromophores which are integral parts of the structural units may induce the photo-degradation.

Polyamides

$$-CH_2CH_2CH_2CO-NHCH_2- \xrightarrow{h\nu} -CH_2CH_2CH_2\overset{\cdot}{C}O + \overset{\cdot}{N}HCH_2-$$

and polysulphones

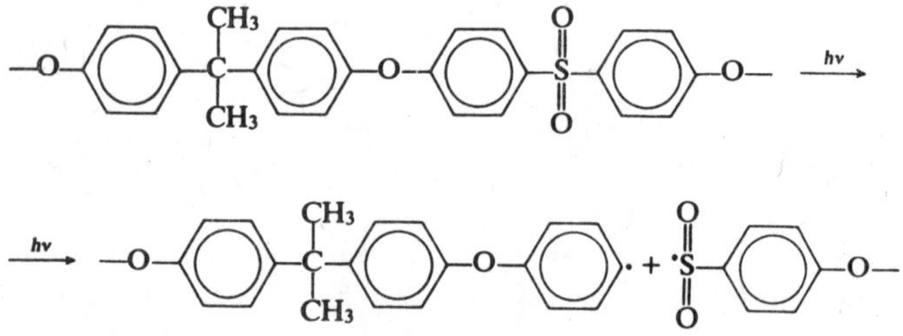

serve as examples.

The same mechanism of photolysis has also been postulated for polycarbonates, poly(ethylene terephthalate) and unsaturated polyesters. Ultraviolet light initiates the dissociation of the C—O bond of polyoxymethylene, polyoxyethylene and polysaccharides

138

Poly(vinyl chloride) is dehydrochlorinated to a system of conjugated double bonds on the main chain. Photo-oxidation of polyurethanes and cellulose is accelerated by the presence of water. Not all reactions of excited states of polymers lead ultimately to degradation, and many polymers may only cross-link. The deliberate promotion of the photo-reactivity of polymers aimed at potential synthetic reactions has been dealt with earlier.

B. THERMOLYSIS OF POLYMERS

The strength of the bonds which may participate in the macromolecular back-bone decreases in the sequence

$$C{\equiv}C > C{=}N \sim C{=}C > P{=}N > N{=}N > Si{-}O > C{-}O > C{-}C > C{-}N >$$
$$> C{-}Si > N{-}O > S{-}S > N{-}N > O{-}O \ (Table \ 7.3).$$ The greater the number of bonds from the left-hand side of this sequence, and the greater the extent of conjugation of multiple bonds, then the more thermally stable is the polymer.

Table 7.3   Dissociation energies of bonds A — B in kJ mol^{-1} [4]

A \ B	H	C	N	O	F	Cl	Br	I	S	Si
H	436	413	391	463	563	432	366	299	339	
C		348	292	352	441	329	276	240	259	290
N			160	222	270	200				
O				139	185	203				369
F					153	254				
Cl						243			250	
Br						219	193	178		
I								151		
S									213	
Si						541	359	289		

C=C 615   C≡N 891   N=N 418
C≡C 812   C=O 716   N≡N 946
C=N 615   N=O 607   O=O 498

The cleavage of the macromolecular backbone may occur in a secondary process following the breaking of bonds in the side groups; the order of the thermal stability of bonds in the side groups is as follows

$$Si{-}F > O{-}H > C{-}F > C{-}H > Si{-}Cl > C{-}Cl > Si{-}Br > C{-}Br > C{-}S$$

139

The magnitude of the dissociation energies of bonds in a macromolecule is thus a first indication in any classification of polymers as thermally stable or unstable.

The average thermal energy of bonds ($\approx kT$) at room temperature (2.5 kJ mol^{-1}) or at higher temperatures (at 1000 °C, 10 kJ mol^{-1}) is however, considerably lower than that for dissociation of the individual bonds in the polymer. The fraction of bonds which reaches the energy equal to the dissociation energy D (T is the temperature in K) is determined by the expression $\exp(-D/RT)$. Thus at 486 °C, in one mol of C—C bonds, only one bond exists having that energy corresponding to its dissociation. Despite this fact, in the temperature interval 350—600 °C, most polymers in an inert atmosphere decompose relatively rapidly into low-molecular fragments. The reason for such behaviour is found in the variable number of weaker bonds always present in the polymer, which start the thermal degradation. As in the case of photo-degradation, these defects may consist of some end-groups of the macromolecule, peroxyl groups or other irregularities in the polymer chain. One might note that one broken bond is found in one mole of peroxyl bonds even at 30 °C.

Until recently, the head-to-head or tail-to-tail configurations of monomer units in a polymer chain were considered as unstable structures, while the head-to-tail structure was regarded as stable. The results of the thermolysis of poly(methyl cinnamate), where head-to-head isomers degrade faster than head-to-tail accord with this view

$$
\left[
\begin{array}{c}
\quad\; \mathrm{C_6H_5}\;\; \mathrm{C_6H_5} \\
\mathrm{-CH-CH-CH-CH-} \\
\quad\; \mathrm{COOCH_3}\qquad\mathrm{COOCH_3}
\end{array}
\right]_m
\;>\;
\left[
\begin{array}{c}
\quad\mathrm{C_6H_5}\qquad\; \mathrm{C_6H_5} \\
\mathrm{-CH-CH-CH-CH-} \\
\qquad\mathrm{COOCH_3}\;\; \mathrm{COOCH_3}
\end{array}
\right]_m
$$

On the other hand, the head-to-head and head-to-tail arrangements of structural units show comparable thermal stability [1] as regards the degradation of polystyrene, polyvinylcyclohexane and poly(methyl methacrylate), expressed by the rate and initial temperature.

A description of the stability of a given polymer usually involves the determination of the rate of thermal degradation at a given temperature. From the temperature dependence of the rate constants, we can then determine the activation energy and pre-exponential factor. For a given polymer structure, the value of the activation energy E of degradation may also be estimated from the dissociation energies D of the newly formed (j) and disappearing bonds (i) such that

$$ E = \Sigma D_i - \alpha \Sigma D_j $$

where α is a coefficient between 0.5 and 1, depending on the type of reaction. Since the degradation of polymers is a complex process, the vaules of α may be determined only approximately and the theoretical values of E, as with the values of D, are only indicative in assessing the stability towards thermal oxidation of a polymer. It may be shown however that the more stable are the products of degradation, the lower is the activation energy and the higher is the rate of reaction, i.e. the higher the exothermicity of a given process, the more probable is its occurrence.

Table 7.4   Data for the pyrolysis of polymers [7].
V_M — monomer yield, E — activation energy,
v_d — rate of decomposition at 350 °C,
T — temperature of a half-life of 30 min

Polymer	V_M/ mol %	E/ kJ mol^{-1}		v_d/ % min^{-1}	T/ °C
Polyoxymethylene	100	42			
Poly(α-methylstyrene)	100	272			
Polytetrafluoroethylene	96	339	344[a]	2×10^6	510
Poly(methyl methacrylate)	95	218			330
Polymethacrylonitrile	85				
Polystyrene	41	230	240[a]	0.235	360
Polychlorotrifluoroethylene	26	239	210[a]	0.200	380
Polyisobutylene	20	205	220[a]	2.400	350
Polybutadiene	20	259			410
Polyoxyethylene	4	193			350
Poly(propylene oxide)	3	84			300
Polymethacrylate	1	143	155[a]	10	330
Polyacrylonitrile	1	243			390
Polypropylene	2	264	255[a] 243[b]	0.069	400
Polyethylene (linear)	0.1	290		0.004	
Polyethylene (branched)	0.025	290		0.008	
Poly(vinyl chloride)	0	134			260
Polytrifluoroethylene	1	220		0.020	
Poly(vinylidene fluoride)	1	200[a]	201[b]	0.020	
Poly(p-xylylene)	0	320		0.020	
Polybenzyl	0	220[a]	210[c]	0.006	
Poly(vinyl acetate)		154[b]			
Poly(vinylidene chloride)		141[b]			
Poly(4-methyl-1-pentene)		222[b]			
Polypyromellitimide		310[c]			
Polytrivinylbenzene		306[c]			
Poly(ε-caprolactam)		180[c]			
Epoxy		126			
Poly(vinyl fluoride)		184[a]	213[c]		

a) reference 5   b) reference 6   c) reference 8

A plot of the experimental values of E (*Table 7.4*) versus the rate of degradation, expressed either by the weight loss at 350 °C or by the temperature associated with a half-life of 30 min., reveals that, although the points are scattered, the general tendency is clear (*Fig. 7.3*).

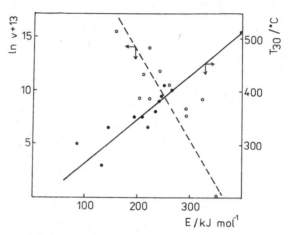

Fig. 7.3. Dependence of the rate v of thermolytic degradation, expressed as the weight loss of polymer per minute and from the temperature T_{30} (i.e. when the half-life for polymer decomposition is 30 min) on the activation energy of the process

(Data from *Table 7.4*.)

The degradation reaction starting on a hydrocarbon chain with side groups R^1, R^2 and R^3 can occur via three possible pathways, depending on the structure of R

142

Bond	$E/\mathrm{kJ\ mol}^{-1}$
$C_2H_5-C_2H_5$	348
$C_4H_9-(CH_2)_3CH_3$	344
$CH_3(CH_2)_2CH_2-H$	411
$(CH_3)_3C-H$	381
$CH_2=CH-H$	440
$CH_2=CH\,CH_2-H$	344
$C_6H_5CH_2-H$	348
$CH_3\dot{C}H\,CH_2-CH_3$	105
$\dot{C}H_2CH_2-H$	159
$\dot{C}H_2-H$	360
$CH_3-\dot{C}=O$	42
$H-CH_2\dot{O}$	101
$C_2H_5-OO\cdot$	117
$C_2H_5-CH_2O\cdot$	50
$(\dot{C}H_3)_2CO\cdot$ $\mid$ CH_3	21

The dissociation energies of the β-bonds in the radicals A, B and C as well as in the unsaturated compound D are lower than in the corresponding saturated compounds (*Table 7.5*) and the subsequent fragmentation will proceed more easily. The secondary radical A gives either a terminal or an internal double bond and the radical B or the low-molecular radical $\cdot R^1$

Radical C prefers the transfer reaction with a neighbouring chain

while radical B fragments mainly to a monomer

143

$$\underset{\underset{R^3}{|}}{\overset{\overset{R^1}{|}}{-CH}}-\underset{\underset{R^3}{|}}{\overset{\overset{R^2}{|}}{C}}-\underset{}{\overset{\overset{R^1}{|}}{CH}}-\overset{\overset{R^2}{|}}{C}\cdot \quad \longrightarrow \quad \underset{\underset{R^3}{|}}{\overset{\overset{R^1}{|}}{-CH}}-\overset{\overset{R^2}{|}}{C}\cdot \; + \; \underset{}{\overset{\overset{R^1}{|}}{CH}}=\overset{\overset{R^2}{|}}{\underset{\underset{R^3}{|}}{C}}$$

and the original structure of the radical B is thus regenerated.

The gradual elimination of monomer molecules from the terminal radical B (the process known as depolymerization) prevails in polymers such as poly(α-methyl styrene, poly(methyl methacrylate), and polymethacrylonitrile, where a radical site of type B is sterically hindered and cannot enter the transfer reaction.

The depolymerization of polytetrafluorethylene is favoured by the very low probability of the transfer reaction of the strongly bound fluorine to sterically hindered alkyl radicals. The displacement of fluorine is, moreover, inhibited by polar factors.

Polyoxymethylene, which has no bulky substituents R^2 and R^3 adjacent to the active centre, depolymerizes from the terminal hydroxyl groups

$$-CH_2OCH_2OH \quad \rightarrow \quad CH_2=O + -CH_2OH$$

but the mechanism of the process is ionic

$$-CH_2OCH_2OH \quad \rightleftharpoons \quad -CH_2OCH_2O^{\ominus} + H^{\oplus} \quad \rightarrow \quad -CH_2OH + CH_2O$$

After the homolytic cleavage of the C—O bond

$$-CH_2OCH_2O- \quad \rightarrow \quad -CH_2O^{\cdot} + {}^{\cdot}CH_2O-$$

alkoxyl radicals will prefer the transfer reaction rather than fragmentation.

The dehydrochlorination of poly(vinyl chloride), the degradation of poly(vinyl acetate), poly(vinyl fluoride), poly(vinylidene chloride) and poly(tert-butyl methacrylate) are examples of the unzipping process which starts with the formation of a double bond on the macromolecular backbone. The only difference from the depolymerization described above is the elimination of compounds derived from the side groups rather than the monomer. Interest has been particularly focused on the *mechanism of the thermal dehydrochlorination of poly(vinyl chloride)* (PVC) which, while being one of the large scale polymers, is thermally the least stable of these. Even at 120 °C, PVC releases HCl and polyene structures are formed on the main chain. The polymer becomes successively yellow, red, brown and finally black. The longest polyene sequence formed on thermal dehydrochlorination involves 25—30 double bonds, with the average value being from 3 to 15 double bonds. The coloration of the degraded samples is due not only to polyene sequences, but also to the presence of carbonium centres. Thus the addition of NH_3 decolours the degraded sample, while an excess of HCl restores the coloration.

The low thermal stability of PVC is a consequence of the cooperative action of several factors [9] such as the presence of some metal cations (Fe, Cu, Co, Zn, Al), certain anions (Cl^-, ClO_4^-, F^- and $B_4O_7^{2-}$), residual initiators (peroxides) some alkaline compounds, double bonds in the structure of the main chain, peroxidic groups incorporated into the polymer during its synthesis and storage, as well as of other structural anomalies (chlorine bound to tertiary carbon, carbonyls in α and β positions with respect to a C=C bond, etc.) [10]. Depending on the manner of activation of the C—Cl bond, the above factors may induce either radical or ionic pathways for dehydrochlorination.

In the case of "ideal" PVC, the initiation step may begin with the cyclic transition state arising from the monomolecular or bimolecular interaction

$$
\begin{array}{ccc}
& & 2\,HCl + \begin{matrix} CH \\ \| \\ CH \end{matrix} + \begin{matrix} CH \\ \| \\ CH \end{matrix} \\
\begin{matrix} CH\!-\!\!-\!H\cdots Cl\!-\!\!-\!CH \\ CH\!-\!\!-\!Cl\cdots H\!-\!\!-\!CH \end{matrix} & & \\
& & HCl + \begin{matrix} CH \\ \| \\ CH \end{matrix} + \begin{matrix} CH\!-\!Cl \\ | \\ CH_2 \end{matrix}
\end{array}
$$

The C=C double bonds which appear in the polymer will promote the subsequent course of dehydrochlorination. At advanced stages of the degradation, the double bonds may react reversibly with HCl or with a suitably oriented segment of the parent macromolecule

$$
\begin{matrix} CH\cdots Cl\!-\!CH \\ \| \quad\;\; : \\ CH\cdots H\!-\!CH \end{matrix} \;\rightleftharpoons\; \begin{matrix} CHCl \\ | \\ CH_2 \end{matrix} + \begin{matrix} CH \\ \| \\ CH \end{matrix}
$$

and the rate of formation of double bonds will continually decrease. The growth of the length of the polyene moiety will thus be interrupted. The above mechanism is supported by the observation of the catalytic effect of HCl on the rate of reaction as well as by the formation of relatively short sequences of double bonds. One should bear in mind however, that in the real polymer, the primary appearance of either radicals or ions may direct the reaction pathway to either the radical or ionic mechanisms and that in the present state of knowledge, this mechanism for the thermal dehydrochlorination is still somewhat controversial.

The thermolysis of cellulose is an example of a combination of both depolymerization and the unzipping of side groups. The destruction of cellulose already starts at 180 °C with the elimination of water which carries on up to 280 °C, at which only dehydrocellulose and a carbonaceous residue are formed [11]. At higher temperatures, cellulose decomposes by a depolymerization reaction to 1,6-anhydro-β-D-glucopyranose (laevoglucosan). The primary alkoxyl radicals

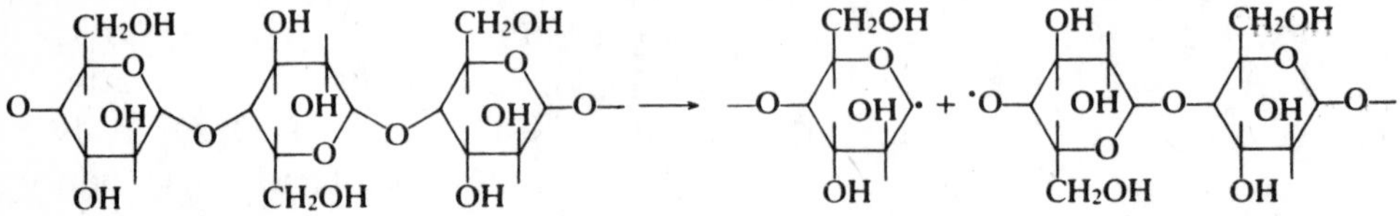

(R˙)

isomerize intramolecularly

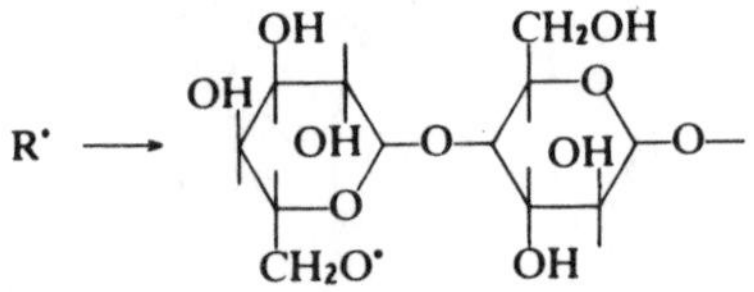

and undergo a subsequent cyclization associated with fragmentation and regeneration of the original structure of the alkoxyl radicals

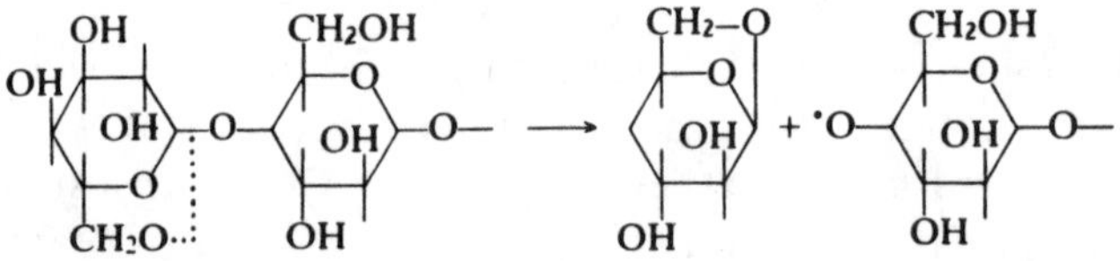

Although many workers adhere to a radical mechanism for the depolymerization of cellulose, similarly to those in the degradation of polyoxymethylene or the dehydration of poly(vinyl alcohol), the ionic pathway is a real possibility.

The thermal stability of a polymer may be increased by a better orientation of the structural units as for highly crystalline polymers. The maximum temperature for the practical application of isotactic polyvinylcyclohexane is, therefore, as high as 220 °C.

By design synthesis, polymers may be prepared with a backbone composed of bonds with higher dissociation energies as in aromatic polyamides [12], ladder and spiro polymers and in other *thermally stable polymers*. Of importance is the increase in the strength of the macromolecular chain on substitution of a simple bond by cyclic units

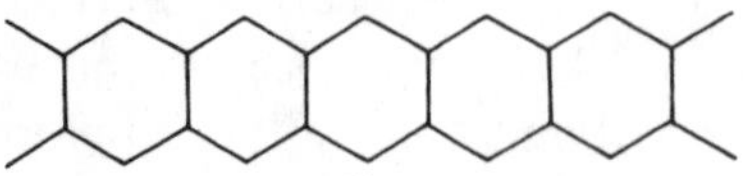

ladder polymer

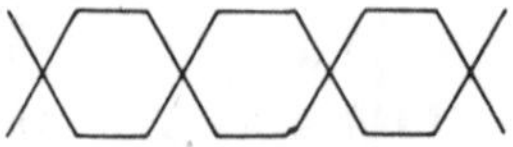

spiro polymer

146

where the individual construction elements involve either aromatic or aliphatic rings including heteroatoms.

The first ladder polymer to be prepared involved thermolysis of polyacrylonitrile

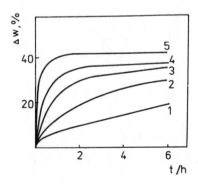

Ladder polymers can be design-synthesized by various condensation reactions of aromatic polyfunctional hydrazides, anhydrides, halides, amines, alcohols and aldehydes. The condensation products have excellent thermal stability and their degradation always yields some carbonaceous residue associated with the simultaneous crosslinking and carbonization of the polymer. The thermal stability of the commonest heteroaromatic polymers decreases in the order polyimides > polybenzoxazoles > polyquinoxalines > polybenztriazoles > poly(N-phenylbenzimidazoles) > polybenzimidazoles.

The degradation of polypyromellitimide of the structure

proceeds appreciably only above 500 °C (*Fig. 7.4*) and the volatile products consist mainly of CO, CO_2, H_2 and H_2O with lesser amounts of HCN, CH_4 and NH_3.

Fig. 7.4. Thermogravimetry curves [8] for the formation of volatiles from polypyromellitimide at 521 (*1*), 542 (*2*), 570 (*3*) 595 (*4*) and 660 (*5*) °C under nitrogen

So-called hybrid polymers such as poly(phenylene siloxanes)

$$\left[\begin{array}{c} R \\ | \\ -Si- \\ | \\ R \end{array} \bigcirc \begin{array}{c} R \\ | \\ SiO- \\ | \\ R \end{array}\right]_n$$

or polycarboranesiloxanes

$$\left[\begin{array}{cc} R & R \\ | & | \\ -SiC(B_{10}H_{10})SiO- \\ | & | \\ R & R \end{array}\right]_n$$

are also thermally stable. They belong to the group intermediate between inorganic and organic polymer systems.

Their extraordinary thermal stability does however, impede the processing of these polymers which is therefore carried out, as for powdered metals, by sintering, or, alternatively, the macromolecular systems are synthesized directly in their final form. A compromise between thermal stability and ease of processing is usually achieved by the designed incorporation of required structural units into the linear macromolecular backbone (*Fig. 7.5*) combined with a complementary crosslinking, as in the case of phenol-formaldehyde resins. The weakest bonds in polymers and the reasons for their appearance have already been referred to in several places. Since most polymer products are in permanent contact with air, the most important reason for the deterioration of the thermal stability of a polymer is the gradual incorporation of labile O—O bonds.

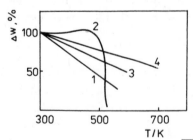

Fig. 7.5. Thermal degradation [1] of polymers of structure $\bigcirc$—X— , where X = CH$_2$

(*1*), CONH (*2*), O (*3*) and CO (*4*)

C. THERMO-OXIDATIVE DEGRADATION OF POLYMERS

The combined effects of heat and oxygen on a polymer (a process referred to as thermo-oxidation or thermo-oxidative degradation), along with photo-oxida-

148

tive degradation are the most frequent processes to which common polymers are subject.

In the presence of oxygen, the temperature of decomposition of most polymers decreases considerably and shifts from 300—600 °C for inert atmospheres to 100—300 °C. Even though O—O groups may penetrate into the macromolecular backbone during the course of a radical polymerization, or during the mechano-processing and moulding of polymers, or via the effect of ozone on double bonds etc., the main pathway of incorporation of O—O groups into the polymer is realized via C—H bonds which are converted to side —OOH groups. In order to increase the thermo-oxidative stability therefore, we need to reduce the concentration of the more reactive tertiary C—H bonds in the polymer, which are the important centres in the propagation stage of radical chain oxidation.

The polysulphone-ether

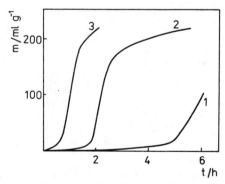

Numerical data are in kJ mol^{-1}

the C—H bonds of which are confined to methyl and phenyl groups belongs, therefore, to those thermoplastic polymers which are very resistant to thermo-oxidative attack [1]. The C—C bonds linking methyl and quaternary carbon, which have a dissociation energy considerably lower than that of the C—H bonds, barely influence the thermo-oxidative stability of the polymer.

Fig. 7.6. Absorption of oxygen [13] during the oxidation of cis-1,4-polyisoprene
1 — 90, 2 — 100, 3 — 110 °C

In its initial stages, in the absence of any initiator, the course of thermo-oxidation is slow. The accumulation of hydroperoxide groups, however, brings about a gradual acceleration of the process (*Fig. 7.6*). The most frequent criterion of the relative oxidizability of polymers, namely *the induction period*, as determined by some convenient method such as absorption of oxygen, depends on the temperature of oxidation (*Table 7.6*). Since the rate constant of

Table 7.6   **Induction periods in the thermal oxidation of some polymers [2]**

Polymer	Induction period/h	
	80 °C	110 °C
Polyethylene (linear)	1 300	70
Polypropylene (atactic)	95	4.5
Polypropylene (isotactic)	130	7.5
Polyvinylcyclohexane	500	300
Polyallylcyclohexane	700	35
Polyallylcyclopentane		1.5
Polystyrene		> 10 000
Polytriphenylpropylene	> 10 000	1 900
Poly(4-phenylbutene-1)	500	30
Poly(5-phenylpentene-1)	360	23
Poly(6-phenylhexene-1)	200	13

the reaction of oxygen with alkyl radicals is extremely high ($\sim 10^8\,dm^3\,mol^{-1}\,s^{-1}$), both the thermo-oxidation and photo-oxidation are usually controlled by the diffusion of oxygen into the sample. If this is so, then the induction period is also some function of the oxygen concentration in the surrounding atmosphere (*Fig. 7.7*). As expected, the activation energy of ther-

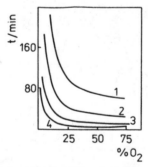

Fig. 7.7. Induction period *t* in the oxidation of polypropylene plotted against concentration of oxygen in the surrounding atmosphere [14]

1 — 120, *2* — 130, *3* — 140, *4* — 150 °C

Table 7.7   **Activation energies for the thermal oxidation of some polymers [6]**

Polymer	$E/\text{kJ mol}^{-1}$
Polyethylene (linear)	137
Poly(vinylidene chloride)	126
Polystyrene	106
Epoxy	92—105
Polypropylene	91—107
Poly(vinyl formal)	88
Poly(vinyl chloride) (suspension)	67—100
Poly(vinyl fluoride)	88
Phenol formaldehyde resins	63—75
Polyacetylene	68

mo-oxidation does not exceed the dissociation energy of the O—O bond (*Table 7.7*).

The thermal decomposition of dialkyl peroxides or alkyl hydroperoxides proceeds with measurable rates usually above 100 °C; some catalysts may initiate the decomposition of hydroperoxides even at room temperature. Some metal ions present in trace quantities in virtually every polymer system may influence the polymer oxidation considerably (*Fig. 7.8*). The approximate

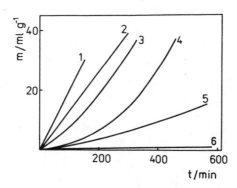

Fig. 7.8. Effect of certain ions on the oxidation of isotactic polypropylene 1 in 1,2,4-trichlorobenzene at 125 °C

Cu (*1*), Cr (*2*), Ni (*3*), Zn (*5*) and V (*6*) $c_{\text{catalyst}} = 7.9 \times 10^{-4}$ mol dm^{-3}. Line (*4*) represents the reference sample of pure polymer

sequence of efficiencies in promoting degradation, which depends on the valence state and type of ligand, is as follows

$$\text{Cu} > \text{Mn} > \text{Fe} > \text{Cr} > \text{Co} > \text{Ni}$$

151

and it can be correlated with the reactivity of these ions (Me) in the reactions of the Haber-Weiss cycle

$$ROOH + Me^{n+} \rightarrow RO^{\cdot} + Me^{(n+1)+} + OH^{-}$$

$$ROOH + Me^{(n+1)+} \rightarrow ROO^{\cdot} + Me^{n+} + H^{+}$$

The mechanism for any particular ion is usually complex. Ions such as Al, Ti, Zn and V decrease the rate of oxidation. The net effect depends significantly on the character of the ligands, and ligand exchange may even turn over the degradation process to one of inhibited degradation.

Although the hydroperoxide mechanism for the autooxidation of polymers is generally valid, there exist examples where the potential hydroperoxide, peroxide or peroxy acid intermediate is so unstable that it cannot be detected in the system. The linear conjugated systems of double bonds in polyacetylene, which are intrinsically thermally stable, react easily with oxygen

$$-CH{=}CH{-}CH{=}CH{-}CH{=}CH{-} + O_2 \rightleftharpoons$$

$$\rightleftharpoons -CH{=}CH{-}\overset{\displaystyle O{-}O^{\cdot}}{\underset{|}{C}H}{-}\overset{\displaystyle \cdot}{C}H{-}CH{=}CH{-} \longrightarrow 2\,{-}CH{=}CHCHO$$

and the polymer decomposes. This oxidation occurs without an induction period (*Fig. 7.9*) [15].

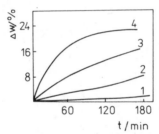

Fig. 7.9. Weight loss of powdered polyacetylene [15] in air
at 55 (*1*), 90 (*2*), 108 (*3*) and 142 (*4*) °C

D. COMBUSTION OF POLYMERS

The glowing and combustion of polymers represents an extreme case of thermo-oxidation. In the surface layer of the burning polymer, the degradation of macromolecules to low molecular products and the formation of solid carbon take place to a considerable extent. In contrast to other degradation processes, a significant role is played here by physical parameters such as heat transfer to the surface, turbulence and diffusion in the gaseous phase, the ratio of surface/ /volume, etc. [16].

152

Chemical reaction in a burning polymer occurs over a steep temperature gradient of up to 1000 K cm^{-1} oriented from the flame to the surface, which affects the hydrodynamic behaviour of the polymer melt on the burning surface. The products of degradation are dispersed into the gas phase and since there is a lack of oxygen in their microdrops, they may act as nucleation centres for the production of soot.

The ignition of a polymer is due either to the direct effect of an outside flame (flash ignition) or to some other thermal source (self ignition) which starts exothermic oxidation reactions at the polymer surface and above the polymer in the gaseous phase. The rate of increase of temperature may be approximated by the equation

$$\frac{dT}{dt} = \alpha A \, e^{-E/RT} w - \beta (T - T_0)$$

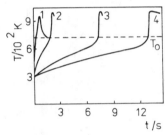

Fig. 7.10. Plot of temperature T against time t based on the energy balance equation for decomposing polypropylene

Parameters $\alpha = 1000$ K, $\beta = 5$ (curve 1), 1 (2), 0.2 (3) and 0.1 (4) s^{-1}. $T_0 = 723$ K, $A = 1 \times 10^7$ s^{-1}, $E = 101.7$ kJ mol^{-1} (from thermogravimetry)

where T is the isotropic temperature of the material, t is the time, $k = A \cdot \exp(-E/RT)$ is the rate constant of the overall degradation reaction and w is the relative residual weight of the sample in relative units, α involves the enthalpy of reaction, β denotes the coefficient of heat transfer and T_0 is the temperature of the heat source. For certain critical values of these parameters, the solution of the above equation may be characterized by an abrupt autoaccelerative increase of temperature. The sample, if in sufficient amount, ignites (*Fig. 7.10*). After stabilization of the flame, the cycle

polymer   ‒ ‒ ‒ ‒ ‒ ‒ ‒ ‒ ‒ ‒ →   volatile flammable products + other products

heat + O$_2$

CO$_2$ + H$_2$O + heat   ←‒‒‒ O$_2$ ‒‒‒   products of cracking

becomes closed.

153

The commonest index of fire risk associated with the use of polymer materials is the *"limiting oxygen index"* (LOI), which expresses the minimum amount of oxygen in its mixture with nitrogen necessary to sustain burning

$$LOI = \frac{[O_2]}{[O_2] + [N_2]}$$

Values of LOI are given in *Table 7.8*. LOI correlates with ΔH_c so, that for most polymers the following relation is valid

$$LOI = \frac{795.3}{\langle \Delta H_c \rangle} \text{ (in \%)}$$

Table 7.8   **Limiting oxygen indices (LOI) and enthalpies of combustion of polymer materials [17]**

Polymer	LOI	$\Delta H_c/kJ\ g^{-1}$
Polyoxymethylene	0.150	17
Poly(ethylene oxide)	0.150	
Poly(methyl methacrylate)	0.173	
Polyethylene	0.175	46
Polypropylene	0.175	46
Polystyrene	0.182	41
Polybutadiene	0.183	45
Poly(vinyl alcohol)	0.225	25
Poly(vinyl fluoride)	0.226	
Poly(3,3-*bis*-chlormethyl oxabutane)	0.232	
Polyamide-6	0.240	31
Polycarbonates	0.227	31
Poly(phenylene oxide)	0.285	
Polysulphones	0.300	
Poly(vinylidene fluoride)	0.437	
Poly(vinyl chloride)	0.470	
Polytetrafluoroethylene	0.950	4

(ΔH_c is the specific heat of combustion expressed in kJ g^{-1}.) The correlation is less satisfactory for polymers which are relatively oxygen-rich such as polyoxymethylene, poly(vinyl alcohol), poly(ethylene terephthalate) and for polymers which yield flame-retarding products such as poly(vinyl chloride) and poly(vinylidene chloride).

An interesting phenomenon was observed during the thermo-oxidation of polypropylene, predegraded polyethylene, polytetrahydrofuran, etc. under conditions when the temperature and oxygen concentration were close to those corresponding to polymer ignition [18]. The oxidation of these polymers is accompanied by the appearance of a cool flame above the polymer surface.

154

Surface oxidation may then proceed as an *oscillatory reaction* and displays regular pulses of temperature, luminescence and pressure (*Fig. 7.11*). The process starts on the polymer surface and propagates through the gaseous phase. The periodical changes of the temperature, luminescence and pressure above the polymer melt corresponds to the final stages of the hydroperoxides in the condensed phase.

Fig. 7.11. Periodical changes of temperature above the surface of molten polypropylene at 300 °C and at an oxygen concentration of 30 mol %

At higher temperatures, hydrocarbons degrade essentially in a pyrolytic way; the reaction of alkyl radicals with oxygen does not proceed via recombination but via disproportionation

$$—CH_2—CH_2\cdot + O_2 \rightarrow HO_2\cdot + —CH=CH_2$$

E. MECHANOCHEMICAL DEGRADATION

When mechanical energy is applied to a polymer during its moulding, milling, mixing and swelling with different solvents, the mutual interpenetration of macromolecular chains leads to the localization of stress on certain bonds with a resulting potential for their cleavage. In such a system, active centres like free radicals are generated. Their subsequent reactions with oxygen or with the surrounding medium, which will depend on the temperature and conditions of diffusion, may have considerable effect on the properties of the product obtained. Scission of bonds is likely to occur within the chain. Hydrocarbon polymers will break down homolytically

$$—CH_2CH_2CH_2— \rightarrow —CH_2CH_2\cdot + \cdot CH_2—$$

while inorganic polymers (silicates, asbestos) and polymers having heteroatoms in their backbone (polysaccharides, proteins) may undergo fission with the formation of polymer ions.

Ultrasonic degradation is regarded as a type of mechanochemical degradation. The effect of the ultrasonic waves on a polymer is indirect, being mediated by the solvent. For degradation to occur, it is necessary that the polymer solutions are saturated by some gas. On application of ultrasound, the microbubbles of the gas oscillate and their adiabatic collapse initiates a pressure wave in their immediate surroundings and an increase in the local temperature which cleave the polymer chains (*Fig. 7.12*).

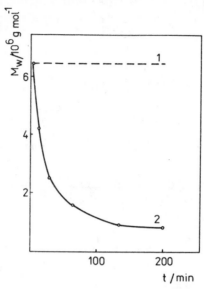

Fig. 7.12. Time dependence of the molecular mass of polypropylene determined viscometrically [1] in toluene at 25 °C after exposure to ultrasonic waves

(1) degassed and *(2)* air saturated solutions. (Power of ultrasound generator 50 W, frequency 0.4 MHz)

The efficiency of chain scission by mechanical energy is relatively low and depends on the possibilities for relaxation of a polymer. When mechanical energy is applied below T_g as in the case of the mechanical drilling of a polymer sample frozen to 77 K in an inert atmosphere, measurable concentrations of free radicals may well be obtained. This method, together with the low-temperature dispersion of polymer solutions in a vibrational mill was one of the first approaches which yielded information on the reactivity of macroradicals. The stage of mechanoinitiation should be distinguished here from subsequent chemical transformations [19].

From the mechanical plasticising of natural rubber in air, it may be seen from the temperature dependence of the molecular mass of the final product (*Fig. 7.13*), that the efficiency of degradation is greater at lower, rather than at higher, temperatures. This is due to the lower plastic deformability of the macromolecular system. On increasing the temperature above 60 °C, the molecular mass of the moulded rubber starts to increase again as a consequence of branching. Further increase of temperature leads mainly to degradation of the macromolecules stimulated by the decomposition of peroxide bonds.

156

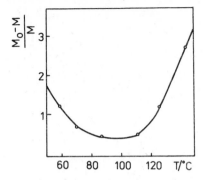

Fig. 7.13. Relative changes in the molecular mass M of natural rubber with temperature after 30 min. of plasticization [20]

(M_0 is the initial molecular mass)

F. DEGRADATION BY OZONE

Photolysis of atmospheric pollutants (nitrogen oxides)

$$NO_2 + h\nu \rightarrow NO + O$$

$$O + O_2 \rightarrow O_3$$

by solar radiation may result in an increase of ozone concentration in certain urban areas and be the cause of a sequence of undesirable reactions occurring not only with synthetic polymers but also with the biopolymers of living organisms. Ozone reacts with practically all organic materials. Although its rate of reaction with alkanes is several orders of magnitude lower than that with alkenes, it may induce their oxidation to such an extent, that the problem is significant.

Ozone may initiate radical reactions as follows

$$PH + O_3 \begin{cases} PO^{\cdot} + HO_2^{\cdot} \\ P^{\cdot} + HO^{\cdot} + O_2 \end{cases}$$

The subsequent fragmentation of the radicals formed may explain the considerable decrease in the molecular mass of certain water-soluble polymers after 4 hr of bubbling their solutions with ozone (*Table 7.9*). The highly degradative effect of ozone may thus be used successfully in the pretreatment of polymer wastes prior to their biodegradation.

Ozone significantly accelerates the dehydrochlorination of poly(vinyl

157

Table 7.9   Changes in the average molecular mass of water-soluble polymers due to the ozone effect [1]

Polymer	Molecular mass/g mol^{-1}		Consumption of ozone/ mg g^{-1}
	original	after ozonization	
Poly(ethylene oxide)	8 000	250	836
Poly(vinyl alcohol)	28 000	460	368
Poly(vinylpyrrolidone)	27 000	560	1 273
Na-Polyacrylate	410 000	250	860
Polyacrylamide	280 000	340	910

chloride); peroxyl groups accumulate in the polymer and its molecular mass falls [21].

The degradative effect of ozone should also be considered in the accelerated photo-oxidation tests when the concentration of ozone may be so high near the ultraviolet light source that it can play the role of a potential initiator of oxidation.

Most interest is, however, focused on the reaction of ozone with the double bonds of elastomers. Elastomers undergoing periodical or permanent stress crack in the presence of ozone. The microcracks formed are oriented perpendicular to the direction of stress. The high reactivity and restricted diffusional mobility of ozone when compared with oxygen means that the reaction of ozone with a solid polymer occurs almost exclusively in the surface layers of the sample.

As pointed out earlier, the reaction of ozone with a double bond starts as a cycloaddition [22]

$$R^1CH{=}CHR^2 + O_3 \longrightarrow R^1CH{-}CHR^2 \text{ (trioxolane ring)}$$

Trioxolanes units are, however, unstable and decompose to aldehydes or ketones and biradicals or zwitterions

$$R^1CH{-}CHR^2 \longrightarrow R^1\dot{C}H{-}O{-}\dot{O} \; (R^1\dot{C}HO{-}O^-) + R^2\overset{O}{\underset{\|}{C}}H$$

The above zwitterions may also be viewed as equilibrium complexes of carbenes $R^1CH{:}$ and oxygen

$$R^1CH^{\oplus}{-}O{-}O^{\ominus} \;\leftrightarrow\; [R^1CH{:} + O_2]$$

The singlet state of the carbene reacts as an ion, while its triplet state gives with oxygen alkylperoxyl radicals.

158

Through the subsequent reactions of these reactive by-products, tetroxanes or branched polymeric peroxides are formed.

$$2\ R^1\overset{+}{C}HO—O^- \longrightarrow R^1CH \begin{matrix} O—O \\ \ \\ O—O \end{matrix} CHR^1$$

$$n\ R^1\overset{+}{C}HO—O^- \longrightarrow R^1CHO—OCH \begin{matrix} O—O \\ \ \\ O—O \end{matrix} R^1\overset{|}{CH} \qquad \begin{matrix} O—O—CH—R_1 \\ \\ O—O—CH \end{matrix} R_1$$

Although the zwitterionic mechanism of ozonolysis has not yet been proved experimentally, it serves as a good illustration of the gradual overlap between radical and ionic reactions.

G. RADIOLYTIC DEGRADATION

The first step in the interaction of highly energetic electromagnetic irradiation with a polymer PH is ionization

$$PH + h\nu \quad \rightarrow \quad \overset{\oplus}{\dot{P}H} + e^{\ominus}$$

The energetic electrons ejected from the polymer cause its further ionization until they are trapped on some group or in a polymer cavity

$$PH + e^{\ominus} \quad \rightarrow \quad \overset{\ominus}{\dot{P}H}$$

or at cation radicals

$$\overset{\oplus}{\dot{P}H} + e^{\ominus} \quad \rightarrow \quad [PH]^*$$

which leads to the excitation of that particular functional group acting as a trap. The excited states of PH and ion radicals are split into radicals, cations and anions.

$$[PH]^* \quad \rightarrow \quad P_1^{\cdot} + P_2^{\cdot}$$

$$\overset{\oplus}{\dot{P}H} \quad \rightarrow \quad P_1^{\cdot} + P_2^{\oplus}$$

$$\overset{\ominus}{\dot{P}H} \quad \rightarrow \quad P_1^{\cdot} + P_2^{\ominus}$$

The radicals in irradiated polymers are evidenced in many experiments. On the other hand, the existence of macroions has been demonstrated directly in only a few cases as in polystyrene, where negative ions were detected [1]. However measurements of the electrical conductivity of irradiated samples and

159

the results of pulse radiolysis indicate their appearance indirectly in certain other cases.

On radiolysis of hydrocarbon polymers, significant amounts of volatile products are formed, especially hydrogen. In the radiolysis of poly(methyl methacrylate) there are also formed CO, CO_2, methane and methyl formate. Each elimination of one side group of poly(methyl methacrylate) is accompanied by β-scission of the backbone

$$
\begin{array}{ccc}
CH_3 & CH_3 & CH_3 \\
| & | & | \\
-C-\dot{C}H-C-CH_2-C- & \longrightarrow & -C-CH=C + \ ^\bullet CH_2-C- \\
| & | & | \\
COOCH_3 \ COOCH_3 \ COOCH_3 & & COOCH_3 \ COOCH_3 \quad COOCH_3
\end{array}
$$

and terminal double bonds and alkyl macroradicals are formed. The alkyl-methylene radical was not, however, traced in the ESR spectra of irradiated samples, a defficiency which was explained by its high reactivity in transfer reactions or by the presence of trace amounts of monomer undergoing subsequent addition reactions. However, the connection between the number of main-chain scissions and of the elimination of side groups which seems to be more generally valid, indicates that such radicals are inherently unstable and eliminate a side group

$$
\begin{array}{cc}
CH_3 & CH_3 \\
| & | \\
-\dot{C}CH_2 & \longrightarrow \quad -C=CH_2 + \ ^\bullet CO_2CH_3 \\
| & \\
COOCH_3 &
\end{array}
$$

The longer the alkyl group in poly(alkyl methacrylates), the higher is the probability of simultaneous crosslinking during polymer irradiation; hexyl represents the last alkyl group where destruction still predominates over cross-linking.

Cellulose is one of the most radiolytically sensitive polymers. Its degradation occurs in both the amorphous and crystalline phases. The rupture of the macromolecular backbone which takes place on the ether C—O bond linking the rings is accompanied by the simultaneous formation of terminal carbonyl groups (20 on average per one broken chain). The latter are formed by the opening and subsequent internal disproportionation of glucopyranose rings.

Degradation prevails over crosslinking during the radiolysis of polyisobuty-lene, poly(vinylidene chloride), poly(α-methyl styrene), polytetrafluoroethylene, poly(vinyl formal), poly(vinyl butyral), polytrifluorochlorethylene, polyoxy-methylene and polypropylene. These polymers lose their mechanical qualities such as tensile strength and elongation even on absorbing low doses of radiation.

The surface modification of polymers in a plasma may also be included in the group of radiolytic degradation reactions. The plasma (mostly of He, Ar, N_2, O_2, CO_2 or H_2) is a gaseous mixture of electrons, ions, excited molecules, free radicals and atoms of a particular gas which comes into contact with the polymer surface and initiates there modification reactions. Most polymers lose some material (*Table 7.10*) during plasma treatment. The plasma of inert gases is used to effect surface crosslingink, while an oxygen or CO_2 plasma for an increases of the hydrophility of polymer surfaces.

Table 7.10 **The rate of weight loss (w) of some polymers [1] after application of a helium plasma at pressure of 13 kPa and a power of 30 W**

Polymer	$w/$ 10^{-3} mg (cm² min)⁻¹
Polyoxymethylene	17.0
Poly(acrylic acid)	16.2
Poly(methacrylic acid)	15.4
Poly(vinylpyrrolidone)	11.9
Poly(vinyl alcohol)	9.4
Poly(ethylene terephthalate)	1.7
Polyethylene	1.2
Polyamide-6	1.1
Polypropylene	0.8

H. IONIC DEGRADATION

Increasing the difference in the electronegativities of the heteroatoms in the backbone or side groups of a macromolecule brings about a shift in the character of the bonds involved to a more polar nature. The ionic course of degradation will be, therefore, more probable for heterochain polymers having ester, ether and acetal bonds, for the C—O and P—O bonds in biopolymers, and the Si—O bonds in polysiloxanes; it is less probable for amide or imide bonds C—N, or C-halogen bonds in side groups [23].

The acid and alkaline hydrolysis of esters or amides is the subject of classical organic chemistry. Ester groups undergo hydrolysis faster than amide groups; the rate depends, of course, on the steric shielding of the bond concerned as well as on the electrostatic repulsion of adjacent ions and on physical parameters such as the tendency towards swelling of the polymer material by a polar solvent and the resulting possibility of ion transport through the polymer.

The acid-catalyzed hydrolysis starts with the attack of a proton (H_3O^+ ion on an ester or amide (X = O or NH) group

161

$$R^1CXR^2 + H^+ \rightleftharpoons R^1\overset{\oplus}{C}X\text{—}R^2 \xrightarrow{H_2O} R^1\text{—}COH + R^2XH + H^+$$

and the intermediate cation undergoes α-scission.

On the other hand, *alkaline hydrolysis* is initiated by reaction of an anion ($^-$OH) with the carbon atom of the carbonyl group

$$R^1CXR^2 + {}^-OH \rightleftharpoons R^1\overset{\ominus}{C}XR^2 \xrightarrow{H_2O} R^1COH + R^2XH + {}^\ominus OH$$

and the anion formed fragments at the β-position. Hydrolytic degradation has a statistical character if all potential hydrolysis sites are equivalent, but it occurs as depolymerization if the primary nucleophilic attack is focused on the end groups of the polymer. Both types of degradation are demonstrated by cellulose which has an ether bond linking the rings which is stable to alkaline compounds but which is destroyed by acids. The attack of protons on the ether oxygen occurs randomly and a mixture of oligosaccharides and cellobiose is formed. Provided that water as the reaction medium is replaced by alcohol in the presence of an acid catalyst, the products have no reducing end groups.

The reaction centres in the basic hydrolysis of cellulose are the end groups in which the carbonyl group is of aldehydic character and thus has reducing properties. The alkaline destruction of cellulose starts then on these groups in a zipper-like manner.

The values of the kinetic parameters for the hydrolysis of the main chain and side groups for some polymers are given in *Table 7.11*. For steric reasons, the

Table 7.11   **Kinetic parameters for the hydrolysis of some polymers [24] (E is the activation energy and A the preexponential factor of the pseudounimolecular reaction)**

Polymer	Medium	$E/$ kJ mol^{-1}	$\log(A/s^{-1})$
Poly(ε-caprolactam)	H$_2$O H$_2$SO$_4$	84	7.1
Cellulose	H$_2$O HCl H$_2$SO$_4$	126	14.2
Poly(N,N-diethylacrylamide)	H$_2$O buffer solution	82	9.5
Polyacrylamide	buffer solution	59	5.6
Poly(vinylpyrrolidone)	buffer solution	105	11.7

hydrolytic scission of side groups is faster than that in the main chain. The steric hindrance of the reaction site may be expressed formally by the 'rule of six', according to which the rate of hydrolysis decreases with an increasing number of atoms in the position 6 with respect to the oxygen atom of the carbonyl group

Table 7.12 **Dependence of the relative rate (v) for the acid-catalyzed hydrolysis of polypeptides on the number (p) of atoms in position 6 with respect to the carbonyl group**

Polypeptide	p	$v^{a)}$
Polyglycine	0	500
Poly(D,L-alanine)	3	100
Poly(D,L-aminobutyric acid)	6	22
Poly(D,L-phenylalanine)	5	13
Poly(D,L-norleucine)	6	17
Poly(D,L-isoleucine)	9	5

a) Related to the rate of hydrolysis of poly(D,L-alanine) in dichloroacetic acid

As regards the scission of peptide (alkylamidic) groups of proteins there exists a fairly good correlation between the number of atoms in the 6-position and the rate of hydrolysis (*Table 7.12*). The rate of hydrolysis may be significantly decreased by the presence of ionized groups such as COOH or NH_2 near the reaction centre, which impede the contact of H^+ or ^-OH ions with the cleaving bond. Acid hydrolysis may also proceed in such a way that the transient cations rearrange primarily to some product which becomes subsequently the subject of a hydrolytic process. It is known that the peptide bonds linking the serine and threonine structural units in silk are very labile in an acid medium. This increased reactivity is due to intramolecular isomerization of the nitrogen cation derived from the peptide bond.

163

The process is, in fact, *N*, *O*-acyl migration. The aminoester formed easily undergoes hydrolytic scission of the main chain.

A similar rearrangement also precedes the hydrolytic destruction of lysozyme, insulin, ribonuclease and glycogen.

I. BIODEGRADATION

In the long-term use of synthetic polymers, degradation due to radiation, oxygen, ozone, water, acids, alkaline compounds and other factors should be considered along with the effects of microorganisms which are mediated by enzymes and microbial metabolites. These effect may become important when using polymeric products in soil or water or in human and veterinary medicine. Such considerations have initiated a research effort towards the synthesis of polymers which are more resistant to microorganisms. Ecological reasons may also require the development of polymers which degrade at rates comparable with those of natural macromolecules. The pathways of degradation of synthetic polymers need to be defined taking into account physiology, biochemistry and the genetics of microorganisms and of microbial assemblies. Since the enzymes produced by microorganisms enable only partial degradation of synthetic polymers, their effect may often be realized after some preceding photochemical, hydrolytic or other type of degradation. As regards the biodegradation of polymers in soil, there may exist a sequence of effects due to particular groups of microorganisms where one group instigates only some part of the total degradation chain [25]. The determination of the biosusceptibility of polymers is associated with considerably larger problems of the reproducibility and reliability of results than found with photo-oxidative or thermo-oxidative degradation. The amount of absorbed oxygen or the rate of growth of the culture deposited on the polymer film are usually measured.

Since the biochemical activity of microorganisms often depends on the presence of trace amino acids, vitamins and growth hormones, this latter test is not always unambigous. The biodegradability of synthetic compounds is strongly limited by molecular sizes [7]. ω-Hydroxycaproic acid is a good source of carbon for microorganisms but its oligomers degrade considerably more slowly. Again, the rate of biodegradation of low-molecular-mass alkanes is much higher than that of polyalkenes. Branched molecules are less susceptible to microbial attack than linear. In hydrocarbon polymers, the end groups are more biosusceptible than the others while in heteroatom polymers biochemical attack is also possible within the chain.

The supermolecular structure of polymers and the compact alignment of semicrystalline polymers markedly hinder the penetration of enzymes to biosusceptible centres, and synthetic polymers degrade microbiologically only with difficulty. The effect of the microorganisms is sometimes weakened by the biocidal character of certain polymer additives which contain Pb, Cu, As, Sn and S or, conversely it may be strengthened by such biosusceptible additives as plasticizers such as sebacates and adipates. The microbiologically most resistant are polytetrafluorethylene, poly(vinyl chloride), polypropylene, poly(phenylene oxide), polystyrene, polyformaldehyde, poly(methyl methacrylate), poly(ethylene terephthalate) and polyamides.

Degradation processes can be divided into desired and undesired. Designed degradation can enable modification of the properties of polymers in order to facilitate their processing. The direct regulation of molecular mass in a polymerization reaction is not always feasible and some polymers are, therefore, processed in a post-degradation reaction. In many cases, selective degradation is tailored towards analytical requirements where the content and sequence of individual mers in copolymers is determined, e. g. by pyrolytic chromatography, the sequencing of proteins and the ozonolysis of double bonds.

The degradation of polymers represents part of a complete cycle encompassing the worlds of inorganic and organic processes and leads to the regeneration of the original natural sources. The current large-scale production of synthetic materials and the consequent increasing quantities of various wastes necessitates the development of ready degradation reactions. The thermal effect is not a negligible factor accompanying controlled combustion. Deliberate degradation evidently has a positive impact on the environment. On the other hand, synthetic and natural polymers age during use and this phenomenon limits their durability. There are various ways to inhibit the ageing of polymers and to optimize their useful life; such optimization is based upon a clear understanding of the degradation mechanisms, which is currently a field of growing importance.

References

1. SCHNABEL, W.: Polymer Degradation, Principles and Practical Applications. Hanser International, Munich 1981.
2. RANBY, B., RABEK, J.: Photodegradation, Photo-oxidation and Photostabilization of Polymers. Wiley Interscience, 1975.
3. GRASSIE, N., WEIR, N. A.: The Photo-oxidation of Polymers, Part I—IV., J. Appl. Polym. Sci., 9, 962—1003, 1965.
4. ČERVINKA O., DĚDEK, V., FERLES, M.: Organic Chemistry (in Czech), SNTL, Prague 1969.
5. DENISOV, E. T.: Kinetics of Homogeneous Chemical Reactions (in Russian), Publishing House Vyssaja Skola, Moscow 1978.
6. DOLEŽEL, B.: The Resistance of Plastic Materials and Rubbers (in Czech), SNTL, Prague 1981.
7. HAWKINS, W. L.: Polymer Stabilization. Wiley Interscience, New York 1972.

8. ANDROVA, N. A., BESSONOV. M. I., LAIUS, L. A., RUDAKOV, A. P.: Polyamides, a New Group of Thermostable Polymers (in Russian), Publishing House Nauka, Leningrad 1968.
   CROSSLAND, B., KNIGHT, G. J., WRIGHT, W. W.: Thermal Degradation of Some Polyimides, British Polym. J., *19*, 291—301, 1987.

9. AYREY, G., HEAD, B. C., POLLER, R. C.: The Thermal Dehydrochlorination and Stabilization of PVC. Macromol. Reviews, *8*, 1—49, 1974.
   GHOSH, P., BHATTACHARYA, A. S., MAITRA, S.: Studies on the Kinetics of Dehydrochlorination of PVC in Solution Induced by a Weak Base, Europ. Polym., J., *23*, 493-496, 1987.

10. BRAUN, D., BOHRINGER, B., IVAN, B., KELEN, T., TUDOS, F.: Structural Defects in Poly(vinyl chloride), Thermal and Photodegradation of Copolymers of Vinyl Chloride with Various Acetylene Derivatives. Europ. Polym. J., *22*, 1—4, 299—304, 1986.
   MARTINEZ, J. A. G., FERNANDEZ, F. G., MANLEY, T. R.: The Initial Thermal Stability of PVC, British Polym., J., *18*, 201—208, 1986.

11. GOLOVA, O. P.: Chemical Effects of Heat on Cellulose, Russ. Chem. Rev., *44*, 687—697, 1975.
    TRASK, B. J., DRAKE, Jr., G. L., MARGAVIO, M. F.: Thermal Properties of Tritylated and Tosylated Cellulose, J. Appl. Polym. Sci., *33*, 2317—2332, 1987.

12. BRAUNSTEINER, E. E., MARK, H. F.: Aromatic Polymers. Macromol. Reviews, *9*, 83—126, 1974.

13. PECH, J.: Synthetic Rubber (in Czech), Publishing House Alfa, Bratislava 1971.
    DEURI, A. S., BHOWMICK, A. K.: Ageing of EPDM Rubber, J. Appl. Polym. Sci., *33*, 2317 —2332, 1987.

14. REICH, L., STIVALA, S. S.: Auto-oxidation of Hydrocarbons and Polyolefins. Kinetics and Mechanisms. Marcel Dekker Inc., New York 1969.

15. WILL, F. C., McKEE, D. W.: Thermal Oxidation of Polyacetylene. J. Polym. Sci., Polymer Chem. Edition, *21*, 3479—3492, 1983.

16. CULLIS, C. F., HIRSCHLER, M. M.: The Combustion of Organic Polymers, Clarendon Press, Oxford 1981.

17. JOHNSON, P. R.: A General Correlation of the Flammability of Natural and Synthetic Polymers. J. Appl. Polym. Sci., *18*, 491—504, 1974.

18. DELFOSSE, L., LUCQUIN, M., BAILLET, C., RYCHLÝ, J.: Upon a New Type of Periodical Reactivity of Heavy Hydrocarbons with Oxygen in the Presence of Their Liquid Phase. Combustion and Flame, *54*, 203—210, 1983.

19. VESELÝ, K.: Mechanochemical Degradation of Polymers (in Czech), Chem. Listy, *68*, 836— 846, 1974.

20. FETTES, E. M.: Chemical Reactions of Polymers. J. Wiley, New York 1964, part I. and II.

21. ABDULLIN, M. I., GATAULLIN, R. F., MINSKER, K. S., KEFELI, A. A., RAZUMOVSKII, S. D., ZAIKOV, G. E.: Effect of Ozone on PVC Degradation. Europ. Pol. J., *14*, 811—816, 1978.

22. RAZUMOVSKII, S. D., RAKOVSKII, S. K., SHOPOV, D. M., ZAIKOV, G. E.: Ozone and its Reactions with Organic compounds (in Russian), Publishing House of the Academy of Sciences of Bulgaria, Sofia, 1983.

23. ZAIKOV, G. E.: Kinetic Study of the Degradation and Stabilization of Polymers, Russ. Chem. Rev., *44*, 833—847, 1975.

24. MOISEEV, J. V., MARKIN, V. S., ZAIKOV, G. E.: Chemical Degradation of Polymers in Aggresive Liquid Media, Russ. Chem. Rev., *45*, 246—266, 1976.
    HOWARD, P., MALOOK, S. U.: Kinetics of Acid Catalysed Degradation of Cellulose Triacetate in Chloroalkane Solvents, Polymer, *28*, 1717—1720, 1987.

25. Ryšavý, P., Brodilová, J., Pospíšil, J.: Biological Systems and their Effect on Polymers (in Czech), Plasty a Kaučuk, *21*, 9—13, 1984.
Zhao, Q., Marchant, R. E., Anderson, J. M., Hiltner, A.: Long Term Biodegradation in vitro of Poly(etherurethane urea): A Mechanical Property Study, Polymer, *28*, 2040—2046, 1987.

VIII. CONTROL OF THE OPERATIONAL LIFE OF MACROMOLECULAR COMPOUNDS

When estimating the operational life of a polymeric material for a particular application, the limiting value of some material property such as tensile strength, elongation, electrical conductivity, permeability to low-molecular--mass compounds, etc. should be established at which the polymer does not fail.

A gradual decrease in the functional properties of a macromolecular compound may occur spontaneously immediately after its production. With regard to the function of the macromolecule in a system this process may be either unwanted or desirable. The latter situation is encountered preponderantly with the biomacromolecules of a living organism, where degradation of the polymeric molecules is an integral part of the system's existence.

The means of promoting analogous behaviour in synthetic polymers are much less selective than in the case of living matter. Shortening the operational life of synthetic and natural macromolecules via accelerated degradation is carried out mostly from the research viewpoint or it is applied to remove plastic materials from the environment.

A much higher practical impact comes from the prolongation of the operational life of macromolecular systems. Apart from modifying a polymer by eliminating its weakest sites or by an increase of its crystallinity, there are other possibilities.

The principal approach involves additives, which protect the polymer by eliminating some of the pathways for generating active sites. Such protective additives or *"stabilizers"* are either added or are inherently present in the polymer, and function by absorbing the energy of an initiation process more efficiently than the polymer itself. Once formed, the primary sites may be deactivated either physically (*by quenching* of the excited states) or chemically (*by inhibiting* chemical reactions). The latter approach is more effective since the interruption of an oxidation chain requires relatively small amounts of the additive although the resulting effect is remarkable.

The species equivalent to the reactive sites which are formed from the protective compound and not on the macromolecule should be less reactive. Only then is the operational life of the polymer may be prolonged.

Since the deterioration of the properties of a polymer is connected with its degradation, the protective compounds or functional groups used to increase its operational life are called *antidegradants* or *stabilizers*. According to how these compounds function as regards a particular type of initiation, we may distinguish photo- and thermo-oxidation stabilizers, flame retardants, antiozonants, antirads (radiation protectants), biological antioxidants, etc. Since the spontaneous degradation of a polymer proceeds under the cooperative influence of oxygen, these compounds and stabilizing systems are normally good antioxidants.

The net effect of using more than one type of additive may be greater or less than resulting from simple addition; the former effect is called *"synergism"* and the latter *"antagonism"*.

A. LIGHT STABILIZERS

Additives which act as light filters, light absorbers and quenchers of excited states are called photostabilizers. Good light stabilization of a polymer requires the prompt elimination of free radicals appearing during photolysis.

Light *filters* form a protective barrier between the light source and the macromolecule. They may be applied as surface coatings or as pigments in the bulk polymer. Their presence may also affect the rate of diffusion of oxygen through the polymer.

Light *absorbers* both absorb and dissipate the energy of ultra-violet light; their effect resembles that of specific filters for the short-wavelength region.

The *quenchers of excited states* deactivate excited states arising from the polymer system and thus inhibit direct scission of the macromolecule.

The choice of a photostabilizer depends on both the polymer system and its potential application; the most frequently used are derivatives of 2-hydroxy-4-alkoxybenzophenone, benztriazoles, salicylates, nickel (II) chelates, carbon black, etc. The stabilizer should be highly light-resistant under illumination; this is the case for aromatic compounds having a hydroxyl group in the 2-position with respect to a particular chromophore, where an intramolecular hydrogen bond is formed

The zwitterion produced on photoexcitation of 2-hydroxybenzophenone, and after transfer of the hydroxylic proton to the oxygen atom of the carbonyl group, isomerizes to the original structure of the stabilizer.

In the reaction sequence, the absorbed quantum is transformed to heat which is harmless under these conditions. In unsubstituted benzophenone, where this cycle of isomerization reactions cannot occur, excited singlet and triplet states are formed, the latter having a reactivity similar to an alkoxyl radical and being able to abstract a hydrogen atom from C—H bonds. Benzophenone is, therefore, a photosensitizer while 2-hydroxybenzophenone is a photostabilizer.

When investigating the effect of individual light stabilizers, a comparison of the rate of free radical accumulation in a polymer after irradiaton with and without the stabilizer may be used [1]. Poly(methylmethacrylate), where most of the radicals formed are converted to one type of terminal macroradical, is one of the most suitable polymers for such a purpose. The experiment is organized

Table 8.1 The effect of additives on the value of p determined from the accumulation of free radicals in poly(methyl methacrylate) after irradiation by light of wavelength $\lambda = 253{,}7$ nm in an atmosphere of helium at the room temperature [1]

Admixture	concentration, %	coefficient of shielding γ	$p = v_0/v\gamma$
2-(2-hydroxy-5-methylphenyl) benzotriazole	0.5	1.46	1.00
	0.8	1.65	0.95
	3.0	4.40	1.00
2-hydroxy-4-methoxybenzophenone	0.3	1.80	1.05
	0.8	4.10	0.75
	1.1	6.30	1.05
2,4-dihydroxybenzophenone	2.0	4.50	0.95
Di-tert-butylphenyl salicylate	2.0	3.10	0.95
Phenyl ester of 3,5-dichlorosalicylic acid	2.0	3.20	0.95
Disalicylate of resorcinol	1.5	3.20	0.95
1,1-dicyano-2,2-diphenylethylene	1.3	1.75	1.10
p-terphenyl	0.3	1.36	0.90
Pentachlorophenyl salicylate	0.2	1.45	0.55
	0.45	1.65	0.45
	1.35	3.75	0.35

so that a stabilized film of the polymer shields an unstabilized one from the light source. The stabilizer effect may be here understood from the value of p determined from the shielding coefficient γ and from the ratio of the rates of free radicals formation without (v_0) and with (v) stabilizer (*Table 8.1*). Provided that $p = 1$, the additive functions as a light absorber; for $p > 1$ other stabilization effects come into play such as inhibition of free radical reactions, quenching, etc; at $p < 1$ the additive acts as a photosensitizer.

In terms of this definition, the commonest stabilizers, the 2-hydroxybenzophenones and the benzotriazoles are absorbers of ultraviolet light, pentachlorophenyl salicylate is a photosensitizer, while *p-tert*-butylphenyl salicylate is both a light absorber and a quencher, etc. The effect of some phenyl salicylates is accentuated by the possibility of their photorearrangement to derivatives of 2-hydroxybenzophenone

The broad mechanism of light stabilization can be determined by a simple experiment [1], where the polymer film loaded by a stabilizer screens the source of radiation from an unstabilized film, and the course of photo-oxidation in both films is compared by measuring the concentration of hydroxyl and carbonyl groups, e.g. by IR spectroscopy. For polypropylene containing some stabilizers, the times of development of a certain optical density of hydroxyl groups are given in *Table 8.2*. The different values for the stabilized and unstabilized films indicate a more complex mechanism of stabilization involving absorption of the light by an additive, quenching of excited states and disruption of the oxidation cycle by a free radical inhibitor or by decomposition of the peroxide. The latter effect is particularly significant for complexes of nickel (II) and phenolic antioxidants, e.g. the widely used nickel β-dibutyl dithiocarbamate.

The combined antioxidant and light-filtering effects may be observed with different carbon blacks, where the efficiency depends strongly on particle size and on the dispersal of the carbon black in the polymer medium. The photostabilizing efficiency of other pigments is usually considerably lower.

The effect of stabilizers operating via only one mechanism can be enhanced by mixing with other additives which suppress the initiation of degradation at all potential stages.

Particularly effective are those mixtures where the individual components show a synergistic effect. The term synergism denotes the prolongation of the

Table 8.2   The light stability ($\lambda = 254$ nm) of stabilized polypropylene films expressed by the time t for attainment of an optical density 0.01 of the band at 3400 cm^{-1} (hydroxyl groups) [1]. (Stabilized films shielding nonstabilized films from the light source)

Additive	Concentration, % w	Time of irradiation of	
		stabilized film/h	nonstabilized film/h
Without additive	—	—	90
Resorcinol monobenzoate	0.1	150	150
4-*tert*-octylphenyl salicylate	0.1	135	135
2(2-hydroxy-3,5-di-isoamyl) benztriazole	0.3	200	210
2-hydroxy-4-dodecyloxybenzophenone	0.2	560	100
Octadecyl-3-(3,5-di-*tert*-butyl 4-hydroxyphenyl) propionate	0.1	440	110
Dimethylglyoxime	0.05	300	100
(Di-isopropyl dithiophosphato) Zn	0.1	380	95
Complexes of Ni			
with: glyoxime	0.1	570	140
dimethyl glyoxime	0.1	650	140
dimethyl dithiocarbamate	0.1	1 310	110
2-hydroxy-4-methoxybenzophenone	0.5	190	95
2,6-di-*tert*-butyl 4-methylphenol	0.1	195	115

lifetime of a principal stabilizer In^{1}H by the chemical or physical effect of a less efficient co-stabilizer In^{2}H in accordance with the scheme

$$R^* + In^1H \quad \rightarrow \quad RH + In^1$$

$$In^1 + In^2H \quad \rightarrow \quad In^1H + In^2$$

R* represents peroxyl or alkyl radicals or electronically excited states of chromophores [2]. Although the presence of an antioxidant in the mixture with a photostabilizer may be highly effective in inhibiting subsequent reactions, we cannot refer to it as synergistic without verification of its antioxidant efficiency at the initiation stage.

A particular photostabilizing efficiency towards polyalkenes and polydienes is displayed by sterically hindered amines (HALS). It has been shown that the efficiency of HALS cannot be interpreted by mechanisms of quenching or absorption but by some chemical process of scavenging peroxyl or alkyl radicals. In this overall reaction, semistable nitroxyl radicals are regenerated periodically [3a] (R = alkyl), as expressed in the following schemes

172

$$O=C \underset{R \quad R}{\overset{R \quad R}{\diagdown N}}\!\!-O^{\cdot} + {}^{\cdot}CH \begin{array}{c} CH_2 \\ | \\ \\ | \\ CH_2 \\ | \end{array} \longrightarrow O=C \underset{R \quad R}{\overset{R \quad R}{\diagdown N}}\!\!-OH + \begin{array}{c} CH \\ \| \\ CH \\ | \\ CH_2 \\ | \end{array}$$

Depending on the structure of the nitroxyl radicals, the above disproportionation reaction may occur in parallel with binding of the nitroxyl moiety to the polymer chain in a termination reaction [3b]

$$O=C \underset{R \quad R}{\overset{R \quad R}{\diagdown N}}\!\!-O^{\cdot} + {}^{\cdot}CH \begin{array}{c} CH_2 \\ | \\ \\ | \\ CH_2 \\ | \end{array} \longrightarrow O=C \underset{R \quad R}{\overset{R \quad R}{\diagdown N}}\!\!-O-CH \begin{array}{c} CH_2 \\ | \\ \\ | \\ CH_2 \\ | \end{array}$$

The nitroxyl radicals are subsequently regenerated in the reactions of either the hydroxylamine or polymeric alkylhydroxylamine with peroxyl radicals

$$O=C \underset{R \quad R}{\overset{R \quad R}{\diagdown N}}\!\!-OH + RO_2^{\cdot} \longrightarrow ROOH + O=C \underset{R \quad R}{\overset{R \quad R}{\diagdown N}}\!\!-O^{\cdot}$$

The variety of structures of oligomeric HALS which have been synthesized enable good compatibility with the polymer and low volatility during processing. Their unexpectedly high efficiency has not yet been explained in every detail. It has been ascertained that HALS associate with hydroperoxides and are consequently always present at the site of an oxidative attack. The complementary stabilizing effect may follow from the reaction of nitroxyl radicals with metal ions of low oxidation state

$$\rangle NO^{\cdot} + Fe^{2+} + H^{+} \quad \rightarrow \quad \rangle NOH + Fe^{3+}$$

which eliminates the harmful decomposition of the hydroperoxides

$$Fe^{2+} + ROOH \quad \rightarrow \quad RO^{\cdot} + Fe^{3+} + {}^{-}OH$$

to reactive alkoxyl radicals.

Chemical synergism is very frequently combined with a physical synergism due to diffusion of the additive from the polymer mass to its surface, where because of the greater concentration of oxygen, degradation is most probable. The polymer mass, where the additive is consumed more slowly, functions therefore as a reservoir of stabilizer.

The branching of a chain oxidation following photochemical initiation may also be interrupted by additives which decompose hydroperoxides in a nonradical way and prevent active free radicals from appearing at all in the propagation stage of the degradation reaction. Such nonradical decompositions of hydroperoxides are associated with organic compounds of sulphur, such as sulphoxides, thiosulphonates, thiocarbamates, and of phosphorus, such as phosphites, etc. or as combinations of both as in dilauryl dithiophosphite or in dilauryl 3,3-thiodipropionate. Most of these compounds are also used as stabilizers of polymers to thermal oxidation.

B. THERMAL AND THERMO-OXIDATION STABILIZERS

Stabilizing additives for polymers which degrade by a radical mechanism usually scavenge reactive free radicals or hydroperoxides from the system. Their effect may be enhanced by simultaneous deactivation of other low-molecular-mass compounds which promote the degradation reaction or by the designed transformation of some part of the polymer molecule to a less reactive centre.

Antioxidants which effect inhibition of the free radical reactions accompanying oxidation of a polymer fall into two main groups.

— *Preventive antioxidants* which impede free radical formation by the nonradical decomposition of hydroperoxides or of other active compounds.

— *Inhibition antioxidants* which intervene directly in the oxidation cycle by fast reactions with alkyl or peroxyl radicals.

In designing inhibitors of radical reactions, one thing should be born in mind, namely that the most effective application of the antioxidant is determined by its lower and upper critical concentration [4] (*Fig. 8.1* and *8.2*). The lower critical concentration of an antioxidant is the minimum amount capable of influencing efficiently the chain oxidation. It is determined by the kinetic length of the chain reaction and thus by the rate of initiation.

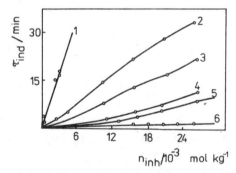

Fig. 8.1. The dependence of the induction periods for the oxidation of polyethylene upon the concentration [InH] of 2,2-methylene-*bis*(4-methyl-6-*tert*-butylphenol)
at 220 (*1*), 230 (*2*), 240 (*3*), 250 (*4*), 260 (*5*), 270 (*6*) °C; oxygen pressure 40 kPa, ref. [4]

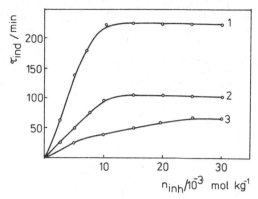

Fig. 8.2. The dependence of the induction periods for the oxidation of polypropylene upon the concentration of 2,2-methylene-*bis*-4-chlor-6-*tert*-butyl phenol

at 180 (*1*), 190 (*2*) and 200 (*3*) °C; oxygen presure kPa. Ref. [4]

Increase of the antioxidant concentration to considerably higher levels causes only slight changes in the induction period and sometimes a pro-oxidation effect can appear. The existence of an upper critical concentration is determined by a concomitant oxidation of the antioxidant; certain products of this type of oxidation, such as the peroxycyclohexadienones produced from phenols, are known as weak initiators of radical reactions, and their effects are evident at higher concentrations of antioxidant.

In selecting an antioxidant for a given polymer, several criteria (*Table 8.3*) should be adopted. Apart from the efficiency of the antioxidant, one must consider its toxicity level, any resulting colour of the polymeric product, its volatility and its compatibility with the polymer, etc.

The most widely used antioxidants are sterically hindered phenols and bis-phenols; other types of molecule are combined with phenols, mostly in synergistic mixtures.

The efficiency of an antioxidant depends naturally on its structure. The correlations which exist between antioxidant efficiency and the standard redox potentials of phenols and amines are sufficiently encouraging to enable the design of new structures with antioxidant properties. A good antioxidant. has a redox potential between 0.7 and 0.9 V.

The sequence of efficiencies of stabilizers determined using low-molecular hydrocarbons may be quite different for polymer systems.

In polymers, the effects of individual stabilizers may be levelled out. This may be due to the specific relation between the polymer medium and the low-molecular-mass compound and to the kinetic inhomogeneity of particular microregions in the solid polymer involving nonhomogeneously dispersed additive.

175

Table 8.3   **Structures of some antioxidants**

Type of compound	Structure
Secondary diarylamines, phenyl-2-naphthylamines, N,N-diphenyl-2-phenylene-diamine, etc.	Ar NH Ar
Products of reaction of secondary diarylamines with ketones and aldehydes	
Primary arylamines	H_2N Ar NH_2
Alkylaryl secondary amines	Ar NH R NH Ar
Sterically hindered phenols	
Sterically hindered bisphenols	
Sterically hindered thiobisphenols	
Polyfunctional phenols	
Nitrogen, phosphorus, and sulphur-containing antioxidants, 2-mercaptobenzimidazole	
(Dimethyl thiocarbamato) Zn, dodecyl 3,3-thiodipropionate	
Alkyl and aryl phosphites, thiophosphites, organic compounds of M(Sn, Ni)	

176

To secure the optimum efficiency of a light-absorbing photostabilizer, a homogeneous dispersion in the polymer is preferred. On the other hand, antioxidants should be concentrated in the amorphous regions of the polymer or on its surface where the oxidative attack commences. The efficiency of stabilizers of low molecular mass may be reduced by their volatility and extractability during processing of the polymer or in its long term use. To suppress such ineffective losses of stabilizer, oligomeric or *macromolecular stabilizers* are synthesized or stabilizing groups are attached chemically to the polymer chain. The resulting reduction in the mobility and capacity of the stabilizing groups to migrate to the polymer surface may, however, lead to a decrease in their stabilizing efficiency, and some optimum degree of motion of an attached group should therefore be achieved. If the attached group is sufficiently mobile, the negative effects of its being attached to the polymer chain are eliminated. 4-(mercaptoacetamido)diphenylamine retains its efficiency as an antioxidant and antiozonant even after attachment to the macromolecules of vulcanized rubber; attempted extraction with hot water lowers it only marginally [5]. The requirement for new stabilizers is avoided when the polymer can be easily modified with a stabilizing group. Low-molecular-mass antioxidants and stabilizers with SH groups can be attached mechanochemically to natural and styrene-butadiene rubbers or to copolymers of acrylonitrile — butadiene — styrene. Such mechanochemical initiation with additives bearing —SH groups may be applied even to polypropylene where the concentration of C=C bonds in the macromolecular backbone is small; thus 1.4 g of 4-(mercaptoacetamido)diphenylamine is bound per 100 g of polymer. This bound stabilizer has an apparently better efficiency than a mixture of (dibutyl dithiocarbamato)nickel (II) and the high-molecular-mass phenolic antioxidant Irganox

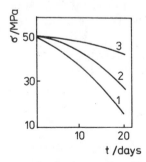

Fig. 8.3. Decrease in the tensile strength of polypropylene film after exposure to hot water and air [5] (90 °C)

1/ reference sample,
2/ sample with high molecular phenol (Irganox 1076)
   and (dibutyl dithiocarbamato)Ni (II),
3/ mechanochemically bound stabilizer 4-(mercaptoacetamido)diphenylamine
Initial concentration of additive — 3×10^{-3} mol kg^{-1}

1076 (*Fig. 8.3*); the decrease in the strength of the polypropylene film after extraction with hot water in air is considerably lower.

Also noteworthy are some quinones or stable galvinoxyl radicals [6] which, in common with nitroxyl radicals, undergo disproportionation with alkyl radicals. The occurrence of a donor-acceptor mechanism of hydrogen transfer between galvinoxyl and polymer radicals can be illustrated as follows

$$PO_2^. + GH \rightarrow G^. + POOH$$

where t-Bu denotes *tert*-butyl, X is a side-group in the polymer chain, G is galvinoxyl radical, GH is reduced galvinoxyl and $P^.$ and $PO_2^.$ are polymeric alkyl and alkylperoxyl radicals. Galvinoxyl radicals convert alkyl radicals to alkenes and become converted to reduced galvinoxyl, which reacts with alkylperoxyl radicals to reform galvinoxyl radicals. The concentration of C=C bonds in polypropylene relative to the initial concentration of galvinoxyl radicals at 200 °C is about 50, which is a markedly higher value than obtained when the polymer melt is stabilized by phenolic antioxidants ≤ 2.

Deactivation of certain metal ions is a process finding wide practical application; these compounds have usually negligible chain-breaking properties and are mostly used in synergistic mixtures with phenolic antioxidants. An example of a metal-deactivator, is the chelating compound disalicylidentetramethylenediamine which deactivates Cu ions but activates Fe and Co ions

N, N′, N″, N‴-tetrasalicylidene(aminomethyl)methane and 1,8-*bis*-salicylidene-3,6-dithiaoctane effectively mask Cu, Fe, Ni, Co and Mn ions. If all the coordination sites of the metal ions are saturated then they do not decompose the intermediate hydroperoxides by a radical mechanism.

178

C. FLAME RETARDANTS

The suppression of polymer combustibility, which has the overall character of retardation, involves chemically or physically active additives or functional groups. Coatings, with the capacity to swell to a charred foam on exposure to heat, prevent access of oxygen to undecomposed polymer and decrease the transfer of heat to its surface, both of which lead to a less flammable material. Another physical effect is exhibited by thermally unstable additives such as $Al(OH)_3$, $Mg(OH)_2$, $CaCO_3$, etc. which, in their endothermic decomposition, liberate nonflammable gaseous products (CO_2, H_2O vapour). The heat consumed in the decomposition of the additive and the dilution of oxygen by the gaseous products combine to shift the thermal balance towards extinction. The additives used as flame retardants can also intervene in the flame chemistry by reacting with reactive intermediates in both the gaseous and condensed phases [7].

Due to the increased temperature (above 300 °C), most thermo-oxidation stabilizers lose their capability to interrupt efficiently the oxidation chains. The high rate of initiation of the decomposition reaction shortens the kinetic length of the chain oxidation so that a higher concentration of antioxidant is required to obtain the appropriate level of inhibition. However, above the upper critical value, a pro-oxidation effect of an additive may be achieved. To inhibit the flammability of a polymer directly in the condensed phase, additive or functional groups are used which promote ready formation of a carbonaceous residue which protects the rest of the material against gaseous exothermic reactions.

The main elementary chemical reactions which should be eliminated by a retarding compound are those of hydrogen and oxygen atoms, and of hydroxyl, alkyl and peroxyl radicals taking place in the gas phase [8]

$$C + {}^{\cdot}OH \rightarrow CO + H^{\cdot}$$
$$CO + {}^{\cdot}OH \rightarrow CO_2 + H^{\cdot}$$
$$H^{\cdot} + O_2 \rightarrow HO^{\cdot} + {}^{\cdot}O^{\cdot}$$

These intermediates may be eliminated through reactions with hydrogen halides, alkyl halides, organic halides of Sb, P, etc. via such processes as

$$^{\cdot}OH + HBr \rightarrow H_2O + Br^{\cdot}$$
$$H^{\cdot} + HBr \rightarrow H_2 + Br^{\cdot}$$

where the bromine atoms produced are incapable of engaging in branching reactions, unlike $\dot{H}$ or $^{\cdot}OH$ radicals.

Due to the increased concentration of carbon-rich intermediates in the flame, one effect of adding halogenated compounds is the increased production of smoke and soot.

The additives which are precursors of the chemically-acting flame retardant should be used in relatively high concentrations to secure a marked retardation effect (about 20 % by weight). Additives containing combinations of halogens, phosphorus and antimony usually reduce the light stability of the polymer and, on their decomposition, produce increased amounts of toxic and corrosive products. Taking into account the above restrictions, the number of practical flame retardants is limited to certain types, such as decabromodiphenyl oxide, triaryl phosphates, Sb_2O_3 + organic halides and additives operating via a physical effect ($Al_2O_3 . 3 H_2O$, borax, H_3BO_3, zinc borates, etc.) [9].

The additive content may be considerably reduced when the volatility and stability of the flame retardant is closely matched with the decomposition and "volatilization" of the polymer. Sb_2O_3, for example, reacts with halogenated compounds in the condensed phase to form volatile $SbCl_3$, which in the gas phase is converted back to antimony oxides by reaction with hydroxyl radicals or with water vapour. Not yet used in practice in the flame retardancy of polymers is the "in situ" introduction of particles with a large active surface which catalyze the termination reactions of radicals in the flame zone. Because of the complexity of the flame system and the lack of reliable correlation between laboratory and large scale tests, an effective solution for the suppression of the flammability of polymers is still under investigation.

D. ANTIOZONANTS

The mechanism of antiozonant action has not been elucidated and antiozonants are selected rather empirically for their purpose. Antioxidants which are effective in radical-terminating reaction only rarely show antiozonant properties; indeed some of them are harmful towards antiozonants. When rubber products are used under essentially static conditions, then satisfactory protection against ozone may be obtained with mixtures of microcrystalline paraffins or ceresins, which diffuse into the surface layers of the polymer and produce a protective solid film. Under conditions of dynamic stress, however, such protection against ozone is unreliable and chemically-acting additives should be used. The commonest of these are aromatic and alkylaromatic secondary amines such as N-isopropyl-N-phenyl-p-phenylenediamine, N, N-dioctyl-p-phenylene diamine and derivatives of tetrahydroquinoline where R is aryl and R′ is hydrogen or an alkoxyl group. Secondary amines may induce the decomposition of the ozonides and initiate crosslinking of the surface layers of a polymer; the antiozonant effect may also follow from direct reaction of the amine with ozone which will diminish the latter's reaction with the double bonds of the elastomer. In the zwitterionic mechanism of ozonolysis, amines may also be involved in reaction with the bipolar ions

$$
\begin{array}{c}
-CH_2-CH^{\pm}-O-O^- \\
+ \\
R^1 \\
\quad \diagdown \\
\quad\quad N-H \\
R^2 \diagup
\end{array}
\longrightarrow
\left[
\begin{array}{c}
-CH_2-CH-OOH \\
| \\
N \\
R^1 \diagup \diagdown R^2
\end{array}
\right]
\longrightarrow
$$

$$
\longrightarrow
\begin{array}{c}
R^1 \\
\diagdown \\
\quad N-OH \; + \; -CH_2-CH{=}O \\
\diagup \\
R^2
\end{array}
$$

which reduces the formation of peroxy bonds in the polydiene and restricts the course of the radical chain oxidation initiated by labile peroxides. One mole of N, N-dioctyl-p-phenylenediamine may thus eliminate 4 moles of ozone, and derivatives of benzidine remove 11 moles, while p-aminophenol accounts for only 2 moles of ozone. The resistance of vulcanized rubber, however, cannot be increased indefinitely since antiozonants also display an upper critical concentration above which they lose their efficiency. This critical concentration depends on the solubility of the antiozonant in the polydiene and on the reactivity of the intermediates formed in the oxidation process.

E. PROTECTION AGAINST OTHER TYPES OF DEGRADATION

The rate of ionic degradation can be reduced by additives which neutralize the catalyst of this undesirable process. Alkaline compounds slow down acid-catalyzed hydrolysis and vice versa. A similar effect on the hydrolytic degradation of polymers is given by compounds reacting with water, such as derivatives of azomethine

$$C_6H_5-N{=}C{=}N-C_6H_5, \text{ etc.}$$

The mechanodegradation of polymers, which is reduced by inhibitors of free radical reactions, is also decreased by low-molecular-mass plasticizers.

Synthetic and natural polymers can be protected against biodegradation by different pesticides, insecticides, fungicides and rodenticides. Compounds offering radiation protection (antirads) are used to prolong the operational life of polymers exposed to ionization radiation. While they are of similar structure to anti-

181

oxidants or photostabilizers, the detailed mechanism of their action is different. Also noteworthy are *the biological antioxidants* such as certain vitamins (vitamin E), phosphatides and compounds produced on cooking food (melanoidines). A strong antioxidant effect is also shown by streptomycin. In living organisms, antioxidants decrease the probability of harmful radical reactions of metabolites such as H_2O_2, etc. It is of interest that antioxidants inhibit the growth of tumour cells (*Fig. 8.4*). This accords with the increase of antioxidant efficiency of the lipid component of membranes before the appearance of a tumour and with the mobilization of the protective functions of an organism [10].

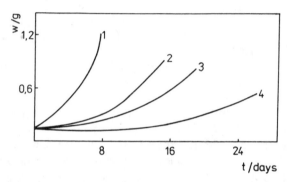

Fig. 8.4. Retardation of the mass increase *w* of leucotic spleen of a mouse by some chemotherapeutic agents [7]

1/ Reference experiment giving the malignant growth of spleen. The growth after the application of:
2/ 2,6-di-*tert*-butyl-4-methylphenol,
3/ 1-(2-chlorethyl)-3-(2,6-dihydroxy-3-piperidine)-1-nitrosyl urea,
4/ 1,2-*bis*-diazoacetylethane

A direct connection between antioxidant effect and harmful mutations in the organism is not, however, obvious and if there is some positive influence, it is likely to be due to elimination of metabolites contributing to the growth of the tumour.

Antioxidants have also been utilized in gerontological experiments oriented towards the prolongation of human life. In spite of some positive results, such chemotherapeutic experiments have been of an empirical nature rather than rationalized theoretically.

References

1. SHLYAPINTOKH, V. Ya.: Photochemical Transformations and Stabilization of Polymers, (in Russian). Publishing House Chimia, Moscow 1979.
2. POSPÍŠIL, J.: Antioxidants, (in Czech). Publishing House Academia, Prague 1968.

3a. SHLYAPINTOKH, V. Ya., IVANOV, V. B.: Antioxidant Action of Sterically Hindered Amines and Related Compounds, Developments in Polymer Stabilization. Editor Scott, G., Appl. Sci. Publishers, London 1982, p. 41—70.

3b. SEDLÁŘ, J., MARCHAL, J., PETRŮJ, J.: Photostabilizing Mechanisms of HALS. A Critical Review. Polymer Photochemistry 2, 175—205, 1982.

4. SHLYAPNIKOV, Yu. A.: Critical Antioxidant Concentration Phenomena and their Applications. Developments in Polymer Stabilization 5. (Editor) Scott, G., Appl. Sci. Publishers, London 1982, p. 1—22.

5. RAAB, M., POSPÍŠIL, J.: Mechanochemical Preparation of Stabilizers bound to Polymer Chains (in Czech). Plasty a Kaučuk 21, 122—123, 1984.

6. SCOTT, G.: Recent Trends in the Stabilization of Polymers. British Polymer J. 15, 208—223, 1983.
Al-MALAIKA, S., SCOTT, G.: Thermal Stabilisation of Polyolefins. Degradation and Stabilisation of Polyolefins, (Editor) Allen, N. S., Appl. Sci. Publishers. London 1983, p. 247.

7. KOŠÍK, M., RYCHLÝ, J., ŠPILDA, I., REISER, V., BAKOŠ, D., BALOG, K.: Polymer Materials and Protection against Fire (in Slovak), Publishing House Alfa, Bratislava 1986.
JENKNER, H.: Flame Retardants for Thermoplastics. Plastics Additives Handbook, (Editors) Gächter, R., Müller, H., Hanser Publishers, New York 1985.

8. CULLIS, C. F., HIRSCHLER, M. M.: The Combustion of Organic Polymers, Clarendon Press, Oxford 1981.

9. ASEEVA, R. M., ZAIKOV, G. E.: Combustion of Polymers, Hanser Publishers, New York 1986.

10. EMANUEL, N. M.: Chemical and Biological Kinetics, Russ. Chem. Rev., 50, 901—947, 1981.

IX. NOVEL PROPERTIES OF MODIFIED POLYMERS

Appropriate adjustment of particular properties of a polymeric material can be made either by the initial selection of monomers or by the organization of the structural elements of the macromolecules. In the first approach, the type of monomer (or monomer mixture) is selected in order to synthesize the most suitable polymer or copolymer. Indirectly, the subsequent modification reactions of the side-groups of chains also belong to this category. These reactions result in copolymers with a sequence of structural units different from that given on the copolymerization of the corresponding monomers. The structural units of various sizes may be used at the organization of the structural elements of the macromolecules; it could relate to the spatial distribution of mers along the chain, regularity in the packing of the macromolecular segments in the crystallite, or the aggregation of chains of one given type in a continuum of chains of another type as in block- and graft copolymers and polymer blends.

It is clear that efforts to obtain new properties can be aimed at various levels in the structure of macromolecular substances. The unique feature of the macromolecules of synthetic polymers and of biopolymers is their dimensional anisotropy, resulting from their essential linearity. The type and sequence of the repeating units in the macromolecular chain is denoted as *the primary structure*. (Such terminology and classification of the various levels of structure while customary for biopolymers, can be applied quite generally.)

The spatial configurations of the mers relative to their nearest neighbours is termed the *secondary structure*. The overall shape of a macromolecule and its supermolecular arrangement are referred to as the *tertiary* and *quaternary structures*, respectively. A transformation of the structure at a higher level can occur without a simultaneous structural change at the lower levels. However, the reverse does not apply and any structural variations at the lower level determine the higher structures in the hierarchy. The degree of the transformation so induced depends on the extent of the changes at the lower levels and on the ability of the higher structural organization to accommodate the structural variations. The chemical reaction of a polymer always represents a change in its primary structure which is subsequently transmitted into changes in the structural organization and of the properties of modified polymers.

Elucidation of structure-property relationships in modified polymers is hindered by imprecise structural data on irregularly-substituted macromolecules. Frequently, a polymer system is described only by the procedure for its preparation rather than by rigorous data on the structure of the macromolecular assembly. Unresolved questions persist even in such a simple reaction as cross-linking, for example, concerning the regularity in the distribution of crosslinks and their character.

However, our limited understanding of the structure of modified polymers in practice derives not only from their complexity; it is in part a consequence of the fact that in the established procedure, synthesis-structure-properties-applications, the second and third steps are frequently omitted in practice and are explored only subsequently when the applications of a modified polymeric material develops. So far little attention has been paid systematically to how the properties of a polymer depend on its method of modification.

Finally, we should point out, that in addition to designed transformations, the structure and properties of macromolecules are affected by degradation and ageing. The question of the stability of properties is pivotal for any long-term use of polymeric materials. So far, in spite of intensive research, our knowledge of the mechanism of the wear and failure of materials is unsatisfactory and limited to generalizations.

Despite these difficulties there already exist some useful rules and correlations between the structure and properties of modified polymers, some of which will be presented in the following sections.

A. TRANSFORMATION REACTIONS AND CHANGES IN MACROMOLECULAR ORGANIZATION

Since the solid-state properties are determined by the structure, the question of the influence of reactions on the resulting properties of polymers can be viewed as a problem of their influence on the structural organization. The organization of macromolecules in the solid state depends on several factors which can be divided into three groups:

— chain length (molecular mass), polydispersity, symmetry and topology (branching) of chains,

— intramolecular conformational flexibility of chains, which depends on internal rotation about the single bonds in the backbone,

— interchain cohesion energy determined by the polarity of the mers and by the influence of the size of the side-groups in hindering the packing of chains.

Chemical reactions affect in various ways these several factors. For example, the degradation of macromolecules mainly alters the first factor. Substitution reactions of polymers changes not only the type and number of side groups but also the flexibility and cohesion of the chains.

Variations in the structure of modified polymers are reflected in some cases in minor changes in their properties, but in other cases these may bring about a striking effect as, for example, with crosslinking when the solubility is eliminated and new elastic properties appear. An alteration in the original structure of the chain always implies a change in its informational content. Note that a macromolecule can be an ideal medium for information storage due to the immense number of possible combinations in its formulation.

a. Crystallinity

Regularity in the organization of macromolecular chains is an important physical parameter which influences especially the mechanical strength and chemical and thermal resistance of polymers. Crystallinity depends mainly on the symmetry of the elements of a polymer chain; the more they are symmetrical, the easier it is for the linear macromolecules to crystallize. This rule applies to stable as well as labile crystallites. The former are typical for the engineering thermoplastics while the latter are formed temporarily in an assembly of stereoregular chains of polydienes which are under stress.

Chemical reactions perturb the original structure of the mers and the symmetry of the initial state controls what changes of crystallinity may take place (*Fig. 9.1*). When a polymer with a regular structure of repeat units reacts, its

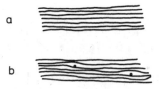

Fig. 9.1. Influence of a chemical reaction on the crystallinity of macromolecular chains
(a) regular arrangement of chains in the crystallite, (b) disordered organization of chains with substituents

crystallinity is reduced during reaction. However, in order to increase the crystallinity of a polymer, it is insufficient to start from an irregular structure of the mers; we require a reaction in which the structural irregularities in the reacting polymer are smoothed out. The chlorination of polyethylene in solution illustrates these principles (*Table 9.1*). The melting temperature of the crystallites is reduced in the initial phase of chlorination since the methylene groups exhibit a higher symmetry than chloromethylene groups; the formation of the latter gradually impairs the regions of regular organization of polyethylene

186

lites

wt. % Cl	mol % CHCl[a]	mol % CCl$_2$[a]	T_m /°C
0	0	0	115
2	2.70	2.34	82
8	10.8	9.37	69
25	33.8	29.3	< 20
40	54.1	46.8	20
45	60.8	52.7	30
54	73.0	63.2	52
60	81.1	70.3	67

a) not determined directly, but calculated assuming the sole existence of these units in the chain; because of the presence of two types of
units, the real values are much lower

chains. However, when the chlorination of poly(vinyl chloride) or partially
chlorinated polyethylene is carried out, the melting point of the crystallites is
increased due to the increased symmetry of the macromolecule induced by the
formation of CCl_2 units.

Increase in the number of defects reduces the crystallinity to various degrees
in different reactions. A minimal number of chemically-induced defects is
needed, especially when crosslinks are formed between macromolecules while
they are in the amorphous state. The linkage of the macromolecules impedes
their independent translation and a reduction in the crystallinity is apparent
when an average of one crosslink is formed per macromolecule. However, an
equivalent network density does not guarantee the same effect on the melting
temperature when crosslinks are formed in the amorphous phase of semicrystalline polymers. In this case the reduction of crystallinity is minimal and even the
opposite trend may be observed. Increase in the melting temperature of polyethylene crystallites after crosslinking in the amorphous phase is explained by
a decrease in the mobility of the polymer chains on melting. The melting
temperature is related to the enthalpy and entropy of melting by the relation

$$T_m = \Delta H_m / \Delta S_m$$

The number of possible conformations of the polymer chains, and hence their
entropy, increases on melting. The conformational flexibility of the chains in the
amorphous state is diminished by crosslinking; the entropy change is thus
reduced and the melting temperature increases.

The reduction of the changes in conformational entropy on melting and its
effect on T_m of polyethylene crystallites can be demonstrated by heterogeneous
chlorination and sulphonation [1]. Under heterogeneous conditions, hydrogen

atoms are initially substituted at the defect sites of the crystallites and in the amorphous intercrystallite regions. This heterogeneous distribution of the newly-bound substituents affects T_m much less than the analogous fully-random substitution along the chain. Thus, the melting point decreases on heterogeneous chlorination by only a few degrees whereas in statistical chlorination this decrease amounts to 85 °C.

The specific effect of heterogeneously substituted chlorine on T_m is associated with a change in the mobility of the chains in the amorphous parts of the macromolecule and, to a lesser degree, also with the increase of the number of defects in the crystallite. The reduced mobility of the chains in the amorphous phase, characterized by the enhanced value of T_g, stabilizes the arrangement of the macromolecules in the crystalline domains and thus increases their melting temperature. An increase in T_m above the value for the non-chlorinated sample is observed at 40 % (wt) of heterogeneously-bound chlorine. Obviously, the chlorine content in the amorphous parts of the sample is higher than this mean value.

The degree of crystallinity depends not only on how the substituent groups are distributed along the chain but also on the polarity and volume of the substituent. The presence of strong intermolecular forces such as hydrogen bonds or dipole interactions promotes a close approach of the chains which manifests itself in an increase of crystallinity. Such an enhancement of crystallinity is illustrated by the hydrolysis of the completely amorphous poly(vinyl acetate) to the partially crystalline poly(vinyl alcohol).

Linkage of polar groups to chains located in the amorphous regions of semicrystalline polymers increases the hardness of polymers. The stronger intermolecular interactions between the crystallites brings about not only hardness but also lower creep of the modified polymers under load.

Major variations in the crystallinity of polymers are related to the isomerization of the elements in the chain. This category includes the epimerization of atactic polymers and the *cis-trans* isomerization of polydienes (*Table 9.2*). The atactic vinyl polymers have an amorphous structure since in their chains the D- and L- forms of the chain elements are randomly distributed. When a homogeneous arrangement of mers exists, the macromolecules can crystallize; isotactic configurations usually lead to a helical structure whereas in syndiotactic polymers the chain backbone is normally in a zig-zag conformation [3].

Regularity of arrangement can also be impeded by macromolecular entaglement. From this reason, the degree of crystallinity of polytetrafluorethylene increases from 60 % to 95 % after mild degradation by ionizing radiation. Secondary crystallization also takes place in polymers which undergo crosslinking rather than degradation under u. v. irradiation (*Tab. 9.3*). Enhancement in the degradation of the exposed parts of the macromolecule may have the same

Table 9.2   The melting temperatures T_m of crystallites of some stereoisomeric polymers

Polymer		$T_m/°C$
Cis-1,4-polyisoprene		28
Trans-1,4-polyisoprene		74
	i	160
Poly(methyl methacrylate)		
	s	200
	i	171
Polypropylene		
	s	138
	i	225
Polystyrene		
	s	250

i — isotactic. s — syndiotactic

Table 9.3   The effect of the time of photo-oxidation (t) on the degree of crystallinity of polyethylene (PE) on irradiation at 254 nm

The type of PE[1]	% of crystallinity		
t	0 h	10 h	50 h
branched 0.927 g cm^{-3}	61.1	61.8	66.8
medium-pressure 0.935 g cm^{-3}	82.3	82.3	92.2
linear 0.984 g cm^{-3}	78.1	85.2	92.2

1) PE samples are characterized by the initial density

origin as the increase in reactivity of polydienes under mechanical strain. Increasing the proportion of the low-molecular-mass compounds in the degraded polymer may bring about a decrease both in T_g and in the degree of crystallinity.

b. Changes in the Amorphous State

The glass-transition temperature T_g is determined by the mutual interplay of all the factors mentioned above, i.e. cohesion energy, conformational flexibility and length of the macromolecule. At T_g, the free volume in the vicinity of the chain is sufficiently large to enable the motion of the larger segments, with 50 to 100 atoms in the backbone. The chemical reactions of macromolecules may produce a shift in the value of T_g. For example, degradation reactions should always produce a decrease in T_g since this parameter is related to the number-average molecular mass $\bar{M}_n$ as follows

$$T_g = T_g^\infty - K/\bar{M}_n$$

where T_g^∞ corresponds to a polymer with infinitely large $\bar{M}_n$.

Simultaneously, however, the interchain cohesion may alter during degradation because of the appearance of low-molecular-mass degradation products with a polarity different from those of parent polymer. The final properties of the polymer system can be anticipated from the relation between the cohesion energies of nondegraded polymer and of the degradation products. Low-molecular-mass compounds with a polarity similar to that of the polymer brings about plasticization of the polymer and hence a decrease in T_g. The small molecules produced on oxidation associate into heterogeneous microdomains due to their higher cohesion energy and do not influence the value of T_g.

Crosslinking enhances T_g because each crosslink represents an additional restriction to segmental mobility, with a reduction of the free volume. It is well known that the vulcanization of natural rubber produces the material ebonite, which is hard at ambient temperature, and it is found that the value of T_g for crosslinked rubber increases linearly with the sulphur content. Similarly, T_g is increased by about 6 °C for each 10 newly-formed crosslinks per gram of polymer in peroxide-crosslinked polyisoprene. The analogous increase in T_g is 11 °C for polystyrene crosslinked by divinylbenzene and 12 °C for poly(methyl methacrylate) crosslinked with ethylene glycol dimethyl methacrylate. In order to increase the value of T_g in polystyrene by 10 °C it is necessary to form at least one crosslink per each original linear macromolecule. Any additional crosslinks which increase the network density result only in a minor increase in T_g [5, 6].

In branched polymers, the two opposing effects are present; an increase in the number of end-groups in the branched polymer provides a larger free volume whereas the higher number of junctions in the chains reduces the chain mobility. Usually the first factor prevails and a decrease in T_g is observed in branched polymers relative to their linear counterparts.

The final product of transformation reactions is usually a copolymer with a value of T_g different from the parent polymers. Several relations have been proposed to predict T_g for statistical copolymers formed from A and B mers. For example, a linear relation applies to a pair of mers not differing much in flexibility and polarity

$$(T_g - T_g^A)w_A + K(T_g - T_g^B)w_B = 0$$

where w represents the weight fractions of the mers and the T_g's are the glass-transition temperatures of the homopolymers and copolymer; K is a parameter determined by the coefficients of volume expansion of the homo-

190

polymers above and below T_g. The equation describes fairly well the dependence of T_g on copolymer composition, for example for polystyrene copolymers (*Fig. 9.2*), nevertheless, in some cases marked deviations are observed. The direction of the shift in T_g values can be estimated from the anticipated changes in the chain flexibility and interchain cohesion. For example, the introduction of bulky substituents (reduction of flexibility) or of polar groups into a nonpolar chain (increase of cohesion) enhances T_g for a modified polymer relative its parent.

In block copolymers the different zones, each consisting of identical mers, can manifest themselves independently. Depending on the composition of the co-polymer, sometimes two glass temperatures can be observed, i.e. one for each of the two types of block, indicating the presence of a domain structure. Sometimes however only one glass temperature is found as for example in block copolymers of acrylonitrile and methyl methacrylate (*Fig. 9.2*), and the above equation can be used for its prediction. Similarly, the number of T_g's observed in interpenetrating networks or in polymer blends depends on the compatibility of their components and on the resulting morphology.

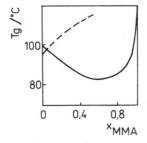

Fig. 9.2. Variation of T_g with composition for the statistical (————) and block (— — — —) copolymers of acrylonitrile and methyl methacrylate (molar fraction X_{MMA})

c. Structure in Solution

Chemical transformations in macromolecules can have some consequences on their spatial organization in solution. In polymer-analogous reactions at either the chain backbone or the substituents and in some reactions involving chain scission, the overall shape of the chain does not change greatly and is represented for synthetic macromolecules by a statistical coil. However, several polymers with semirigid and rigid chains are known which form in solution liquid-crystalline phases. The formation of a liquid-crystalline structure requires, in addition to chain rigidity, that the polymer concentration exceeds some critical in value. Under these conditions, the rod-like macromolecules are aligned predo-

191

minantly in one direction and form the nematic state. Chiral macromolecules may give in solution the chiral (cholesteric) mesophase. The chirality of the mesophase derives from asymmetric interactions between chiral macromolecules and solvent molecules; the macromolecules are oriented in parallel with the helical long-range regularity along the chain axis. Chemical transformations of rod-like macromolecules do not inhibit the formation of the liquid-crystalline state, but simply modify some of its characteristic as shown for example by various derivatives of cellulose [7]. Clearly the process of substitution in this example does not change the overall shape and organization of the chain in solution.

More pronounced changes in the organization of chains in solution are characteristic of modification reactions producing branched and block co-polymers. The differences in chain geometry due to branching are reflected in the variation of the dimensions of the coil in solution in comparison with linear polymers of corresponding chemical composition and of identical molecular mass. Therefore, measurements of the intrinsic viscosity and other properties in dilute solutions are used in the study of the structure of branched polymers [8]. The quantitative determination of the influence of the degree of branching on the intrinsic viscosity is complicated by the variety of ways of branching as shown in *Fig. 4.1*. In contrast to star-branching with only one centre, there are several branching points in comb-like or cascade branching. Moreover, the branches can be of identical or variable length and chemical composition.

The degree of branching (B) of chains with an identical mean number molar mass is usually determined experimentally as the ratio of the intrinsic viscosities of branched (b) and linear (l) macromolecules in a theta solvent

$$B^\epsilon = ([\eta]_b / [\eta]_l) < 1$$

The exponent ϵ lies in the range from 0.5 to 1.5. The difference in intrinsic viscosity between branched and linear macromolecules is reduced when good solvents are used. The magnitude of the exponent depends on the number, rigidity, length-homogeneity and regularity in location of the branches. Chemically homogeneous macromolecules of a given molecular mass exhibit the highest intrinsic viscosity when they are linear. For star-like and cascade-like branched macromolecules the ratio $[\eta]_b / [\eta]_l$ is from 0.5 to 0.75.

Deviations from the random coil structure are found in block copolymers with differing affinity of the components towards the solvent and various ordered structures can be formed in solution. For example macromolecules of natural rubber with poly(methyl methacrylate) side-chains give highly viscous solutions in a solvent common to both polymers. When a non-solvent is added to the solution (by about 25 % of solvent volume), the solution become turbid and its viscosity drastically diminishes. This is associated with additional coiling

192

of those segments the solubility of which has been reduced. The tightly-coiled portions of the macromolecules cannot fully precipitate since they are linked to the branches of the polymer which are well solubilized in the solvent. The solvated polymer acts here as a stabilizer of the precipitated colloidal particles. Thus a dispersion can be prepared from these copolymers which still flows at a dry content of over 40 % (wt).

The domains of tightly-coiled segments of macromolecules in solutions of block- or grafted copolymers form layered, cylindrical or spherical micelles. The supermolecular structure depends on the molar volumes of the tightly-coiled and of the dissolved segments of the copolymer. A lamellar arrangement dominates in A—B block copolymers with insoluble blocks B

```
A   A   A   A   A   A   A
|   |   |   |   |   |   |
B   B   B   B   B   B   B

B   B   B   B   B   B   B
|   |   |   |   |   |   |
A   A   A   A   A   A   A
```

and when the molar volumes of the A and B blocks are comparable. Copolymers with a larger volume of dissolved blocks A form cylindrical or spherical micelles which can be arranged into a hexagonal or cubic structure.

A similar arrangement of macromolecules can also be observed in block copolymers without the solvent. However, the resulting structure depends on the solvent originally present. The nature of the solvent determines the shapes of the chains, which form a continuum between aggregated clusters and other kinds of chain. Thus, the properties of a film obtained from identical block- or graft copolymers is influenced by the solvent used in the sample preparation. When a solvent of poly(methyl methacrylate) is used, the resulting film of grafted natural rubber is hard and tough.

A soft elastic film is formed when a solvent for the rubber segments is used and subsequently evaporated. The elasticity of the film is promoted by the chemically-anchored particles of aggregated (and at ambient temperature immobilized) poly(methyl methacrylate) chains functioning as a filler.

B. SOLUBILITY

The solubility of a polymer is a property which is an sensitive index of the outcome of transformation reactions of macromolecules. The change of Gibbs energy

$$\Delta G_{\text{mix}} = \Delta H_{\text{mix}} - T\Delta S_{\text{mix}}$$

has to be negative for dissolution to occur. Since the entropy term increases in the process, the sign and the magnitude of the enthalpy term should determine the solubility of a macromolecule in a given solvent. This simplified reasoning applies to a solution without specific interactions, such as hydrogen bonds when the components are randomly mixed. The enthalpy of mixing ΔH_{mix} expresses the change in intermolecular interactions on transfer of amorphous polymer into solution.

The solubility parameter defined as

$$\delta = (\Delta E_{\text{vap}}/V)^{1/2}$$

where ΔE_{vap}, the vaporization energy of the compound with molar volume V, is used as a measure of intermolecular cohesion. In general, a polymer dissolves when a chemical and structural similarity exists between its mers and the molecules of solvent, for example, when their polarity is similar. This rule can be quantified by the condition that the solubility parameters of the polymer δ_p and of the solvent δ_s should not differ by more than one or two units when they are expressed in $J^{1/2}\,cm^{-3/2}\,mol^{-1}$. This situation is illustrated in *Fig. 9.3* which shows how, by reducing the difference $\delta_p - \delta_s$, the polymer at first swells and is finally fully miscible with solvent [9]. The solubility increases with reduction in the molecular mass of the polymer. Transformation reactions either on the backbone or at the side-groups can considerably modify the effective value of the solubility parameter δ_p. Apart from changes in polarity, the process of dissolution is also affected by modification-induced changes of crystallinity. It is well known that crystalline polymers have a limited solubility and become soluble at the temperature close to the melting point of the crystallites. Transformation reactions creating defects in the crystallites evidently enhance the solubility of the polymer.

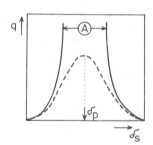

Fig. 9.3. Dependence of the degree of swelling q on the solubility parameter δ_s for linear (————) and crosslinked (-------) polymers. *A* — region of unlimited solubility (miscibility) of the polymer in the solvent

The degree and nature of the substitution in polymer-analogous reactions determines the changes in solubility. The derivatives of cellulose are well-known examples since their solubility is increased by the introduction of substituents into the constituent units which destroys the intermolecular hydrogen bonds in the parent cellulose. The impact on the solubility is diminished by reducing the degree of substitution but can be observed even when the polymer samples differ only in the structure of their end-groups [10].

Miscibility is restricted not only by a very large difference between the solubility parameters of polymer and solvent but also by chain crosslinking (*Fig. 9.3*). The maximal degree of swelling is observed when the cohesion energies of solvent and of polymer are identical. The extent of swelling depends on the density of crosslinks and on the polymer-solvent interaction. Conversely, the crosslink density can be determined from swelling measurements for systems where the parameters for polymer-solvent interaction are available.

The average number n, of crosslinks per macromolecule can be calculated from the solubility of a partially-crosslinked polymer sample with the most probable distribution of molecular masses by the relation

$$n = (s + s^{0.5})^{-1}$$

where s is the weight fraction of the dissolved part.

When macromolecules are destroyed during the crosslinking process, the ratio of destruction and crosslinking p/q can be determined from the amount of the soluble fraction and from the variable amount of radicals (R) formed in the system by the relation

$$s + s^{0.5} = p/q + k/R$$

The solubilities of grafted and linear macromolecules are similar when the chemical compositions of the backbone and of the branches are identical. In other cases, the solubility of branched polymers is always lower and is minimal for copolymers with long-side branches. The larger the difference between the solubility parameters of the backbone segments and the branches, the more difficult it is to find a solvent for the copolymer. The use of a solvent mixture is the most convenient in this case.

C. PERMEABILITY

The diffusion and permeability of low-molecular-mass compounds in the liquid or gaseous state is closely interconnected with the solubility of a polymer. The permeants are adsorbed in the surface layer of the film, diffuse through the membrane and are released on the other side of the membrane. The process in its stationary state is characterized by the coefficient of permeability P, defined as the product of the diffusion coefficient and of the solubility coefficient, $P = D.S$. The coefficient S is defined by Henry's law, i.e. the direct proportionality between the gas pressure and the amount dissolved in the polymer.

The diffusion and permeation of gases, vapours and liquids is important in the preparation and processing of polymers and especially in their use as protective coatings, packings and as textiles. For materials used in these capacities one has to bear in mind that chemical transformations of macromolecules affect the values both of the coefficients of diffusion and of solubility. Nevertheless, to a crude approximation one can state that changes in material structure have a greater effect on coefficient D, whereas the solubility S depends mainly on the character of the low-molecular-mass compound. The permeability is determined by factors identical with those operating in the organization of macromolecules in the solid state and determining the temperatures T_g and T_m [12]. For example, an increase in the number of acetyl groups in cellulose acetate brings about a reduction of interchain cohesion and of crystallinity and thus an increase in the permeability coefficient. Branching has an analogous effect and thus it is not surprising that the permeability of small molecules is higher in branched than in linear polymers (*Table 9.4*).

Table 9.4   The relative permeabilities P of branched (b) and linear (1) polyethylene (PE) towards some low-molecular-mass compounds at 21.1 °C and 133 Pa pressure difference

Compound	Type of PE	P	P_b/P_l
Ethyl acetate	l	1	1.2
	b	1.2	
Heptane	l	3.4	5.8
	b	19.7	
Benzene	l	0.3	107
	b	32.1	
p-Xylene	l	2.7	13
	b	35.4	

To assess the oxidative weathering of a polymer, it is important to establish both its permeability by oxygen and the variation of P in the course of oxidation. *Table 9.5* shows for sodium-butadiene rubber that both the coefficients of diffusion and permeation of oxygen decrease with an increase in the degree of oxidative modification of the polymer. The reduction of the permeability of oxygen following oxidative modification is apparently associated with the reduced conformational mobility of oxidized polydiene. The same reason lies behind the greater permeability of polydienes in comparison with polyalkenes. In agree-

196

Table 9.5

Table 9.5   The variation of the diffusion coefficient D and permeability P in sodium-butadiene rubber at 40 °C with the degree of oxidation X (in mmol of reacted O_2 per mol of mers of rubber)

X	$D/10^6 \text{ cm}^2 \text{s}^{-1}$	$P/10^7 \text{ cm}^2 \text{s}^{-1} \text{ MPa}^{-1}$
0	0.92	8.4
50	1.18	10.4
150	0.21	0.17
250	0.061	0.03
350	0.048	0.03
450	0.044	0.02

ment with this argument, the coefficient P is also reduced by the hydrogenation of the polydienes and by all processes eliminating double bonds in the polymer backbone.

The permeability of gases and vapours is also reduced by the heterogeneous chlorination of polyethylene. Again, this is a consequence of the increase of T_g in the amorphous state of the semicrystalline polymer. The highly elastic material prepared by the bromination of butyl rubber has a low air-permeability particularly which is particularly suitable for high-quality inner tubes. The lower air-permeability derives in this case from the lower solubility of oxygen and nitrogen in brominated polyisobutylene compared with polydiene rubbers.

The permeability of a low-molecular compound in a polymer is determined primarily by the magnitude of the free volume. Hence it is natural that crosslinking, which reduces the segmental mobility and the free volume, also diminishes the coefficient P. Since, as noted in this Chapter, crosslinking enhances T_g, a correlation between T_g and permeability P can be expected, as in *Fig. 9.4* for natural rubber vulcanized with variable amounts of sulphur.

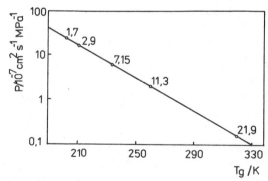

Fig. 9.4. Variation of the permeability P of nitrogen with the glass-transition temperature for vulcanizates of natural rubber with varying degrees of bound sulphur (in wt. %)

A step-like decrease of permeability is usually observed at the transition from the amorphous to the crystalline state. The transformation reactions creating defects in the crystals may facilitate the permeation of small molecules. The permeability of semicrystalline polymers is determined primarily by the amount of the amorphous phase. However, modification of the supermolecular crystalline structure, such as the size of the spherulite, also influences the permeability, since the amorphous component is localized on the periphery of the spherulites where numerous cracks exist due to internal stress. Hence, the coefficient P for nitrogen in polypropylene increases rapidly when the spherulite diameter exceeds 100 μm.

Measurements of permeability are used in the determination of structural changes in modified materials. For example, information on the morphology and microheterogeneity due to phase separation can be obtained for mutually interpenetrating networks.

A common goal in the modification of permeability is the enhancement of the selectivity of polymeric materials towards permeation by small molecules. The selectivity is influenced especially by the grafting of polymers [13] and the resulting materials are able to differentiate fine details in the structure of permeates. Thus the grafting of polyisoprene by poly(methyl methacrylate) enhances its selectivity towards the transport of inert gases. The introduction of about 15 % (wt) of poly(methyl methacrylate) reduces the permeability of helium to 75 % and of argon to 15 % of the original value. Grafting generally affects the transport properties of larger penetrants, whereas for tiny atoms such as helium the free volume needed for migration is barely affected.

D. SURFACE PROPERTIES

Frequently, the wide range of properties required in particular applications is not satisfied by materials based solely on the usually available polymers. Even knowledge of structure-property relationships might be of little help since variation in the chemical structure alters all the properties but to different degrees. In some cases, a special "combination" of polymers may prove suitable, namely, that the core part of the product consists of the original polymer whereas its surface is modified.

Chemical modification affects the surface energy of polymers. The introduction of polar groups into the surface layers of a hydrocarbon polymer enhances its surface tension (*Table 9.6*) and the angle of wetting by polar liquids is reduced. The high-energy surface of modified polymers originates not only from the changes in its chemical structure but also from its more particulate character with an effectively larger area created after reaction. These two factors act concertedly in the modification of nonpolar polymers but their effects may

198

Table 9.6   Surface tension (γ) of some polymers

Polymer	$\gamma/\text{mN m}^{-1}$
Polytetrafluoroethylene	18
Polydimethylsiloxane	24
Poly(vinylidene fluoride)	25
Poly(vinyl fluoride)	28
Polypropylene	29
Polyethylene	31
Polystyrene	33
Poly(methyl methacrylate)	36
Poly(vinyl chloride)	36
Poly(vinyl alcohol)	37
Polycaprolactam	46
Cellulose	50

compensate on the substitution of nonpolar groups in polar polymers. The effect of a rough surface is observed particularly when reactive reagents are used which bring about some degradation of the macromolecules to small molecules. As typical examples, the treatment of polyalkenes by an oxygen plasma or elementary fluorine or of polytetrafluorethylene by sodium, can be mentioned. Partly because of problems in the measurement of surface area, there is a lack of quantitative data on the changes of the surface energy of polymers after modification and the surface properties are evaluated qualitatively from their performance in a particular application.

The surface grafting to a polyalkene fibre by a polar monomer has been investigated intensively for some time. It was motivated by the aim of consolidating the good mechanical properties of polyalkenes with the superior sorptional properties of the surface-grafted polar chains. One has to bear in mind that polyalkenes are much cheaper than the typical fibre-forming polymers and thus the realization of this project would bring considerable economic benefit. This modification by grafting increases the moisture uptake, ease of dyeing, softness and comfort during wear. However, the changes in the properties are unpredictable which makes a serious impediment to the application of this process in the textile industry.

The grafting to cellulose and polyamide fibres of dienes has been more successful. The modified material displays better adhesion to the vulcanized rubber in tyres and the whole composite system in a tyre is strengthened by the enhancement of the interphase joint between the cord with the rubber matrix.

The surface of acetylcellulose fibres is utilized by some producers to reduce electrostatic charging and to improve the comfort during wear of fibres in

199

contact with the skin. The reduction of the surface energy of cellulosic fibres is achieved by esterification with perfluoroaliphatic carboxylic acids and is utilized in the oil-repellant finish of working clothing.

The significance of the modification of the surface properties of polymeric materials can be demonstrated by some other examples. Crosslinked hydrophilic polymers with about 20 % (wt) of water are easily tolerated by living organisms. However, the mechanical strength of these gel materials is insufficient for some applications. Accordingly, hydrophilic monomers such as 2-hydroxyethyl methacrylate, ethylene glycol dimethacrylate and N-vinylpyridine are grafted onto prosthetics made from silicone rubber or other elastomers [14]. In another method, heparin is chemically linked to the silicone-rubber surface in order to diminish blood coagulation on the hydrophobic polymer and thus to prevent thrombosis. The anti-thrombogenic action is ascribed to the polar sulphone groups in the polysaccharide macromolecule of heparin. Similarly, the sulphonation of polyethylene also reduces the level of blood coagulation on its surface since its surface tension is raised to $52 \, mN \, m^{-1}$. Approximately one —SO_3H group per 60 methylene units is formed in the surface layer after a three-hour exposure in concentrated sulphuric acid.

The chemical treatment of the surface improves the properties of the polymer in numerous other applications [15] such as in the metallization of polymers, the enhancement of their photochemical stability [16], their membrane permeability [17] and the strength of adhesive joints. Dehalogenation of polytetrafluorethylene by metallic sodium is used as a pretreatment to the glueing of this chemically and physically inert material which could not otherwise be effectively glued [18]. Surface crosslinking can be used in order to increase the strength of glued joints of polyethylene. It was found [19] that the joint based on polyethylene-epoxide resin, is eight times stronger after irradiation by u.v. light in an inert atmosphere or on exposure to a helium plasma. No surface oxidation was detected by reflectance infrared spectroscopy and from the change of the wetting angle of polar liquids it is therefore assumed that the enhancement of the strength of the joint is due to the improved mechanical strength of the polyethylene surface.

Variations in the structure of the surface layers in polymer particles with a diameter of 1 μm are important in colloid chemistry and biochemistry [20]. However, these problems are not yet well understood, as is the situation in several fields relating to modified polymers.

E. MECHANICAL PROPERTIES

The title implies a wide range of properties but most frequently it refers to the relation between stress and strain under certain conditions [21]. The physical response of a polymeric material to mechanical deformation depends on processes operating at the various levels of macromolecular structure, i.e. it is a response not only of single macromolecules but also of their assemblies where the result is a function of the supermolecular morphology.

Several mechanical parameters are illustrated by the general plot of stress versus the relative elongation at a constant rate of deformation (*Fig. 9.5*). The slope of the curve gives Young's modulus E and the stress at the break-point specifies the strength. The toughness is proportional to the area between the experimental curve and the abscissa.

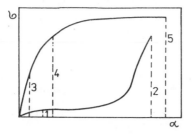

Fig. 9.5. Stress (σ) strain (α) curves for various types of polymer
(*1*) soft and weak, (*2*) soft and tough, (*3*) hard and brittle, (*4*) hard and strong, (*5*) hard and tough

The dependence of the elasticity modulus E on the temperature or on the time-scale of the experiment has a similar shape for polymers differing chemically, although the temperature and the time-scale are naturally different. Accordingly, substitution reactions in the repeat units of the polymer chains considerably affect the modulus. The particular change in the modulus-temperature function greatly depends on whether the substitution reactions conserve or destroy the homogeneity of the macromolecular system, or in other words what type of copolymer is formed (*Fig. 9.6*). Similarly, the modulus-

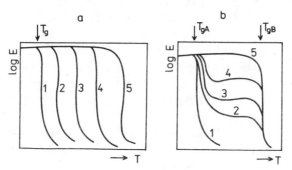

Fig. 9.6. Variation of Young's modulus E with temperature for various copolymers
(*a*) statistical, block or grafted copolymers with demixed phases, (*b*) multiphase (microheterogeneous) block and grafted copolymers,
(*1*) homopolymer A, (*2*), (*3*), (*4*) copolymers with increasing content of B units, (*5*) homopolymer B

201

temperature (or time) function is also affected by crosslinking or by other reactions which change the degree of crystallinity or molecular mass of the polymer (*Fig. 9.7*).

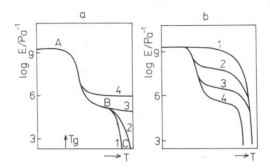

Fig. 9.7. (*a*): Effect of crosslinking and of molecular mass of the amorphous polymer on the dependence of modulus E (in Pa) upon temperature T

(*1*) original polymer, (*2*) polymer of higher molecular mass, (*3*) crosslinked polymer, (*4*) highly crosslinked polymer
The physical states of polymers; A glassy, B rubbery, C viscous
(*b*) The modulus-temperature plot for
(*1*)crystalline (*2*), (*3*) semicrystalline, (*4*) amorphous polymers

Toughness is typical for some copolymers with a microheterogeneous structure. For example, the grafting of polystyrene chains onto a polydiene elastomer brings about a distribution of small rubbery particles in the polystyrene and consequently, the preservation of hardness and strength and an increase in the impact resistance, to give the name "high impact" polystyrene. The rubber particles can absorb and dissipate the energy of impact without the fracture of the product. Similar behaviour is found in block copolymers comprising soft and flexible, and hard and stiff macromolecular segments, forming a two-phase system. The organization of the macromolecular structure depends on the arrangement of the component units in the block copolymer; maximum toughness is observed in block copolymers of the type ABA and (AB)$_n$. In three-block copolymers, it is important that, the terminal blocks have a value of T_g above, and the macromolecular segments B a value below the temperature of application. The hard segments aggregate into evenly distributed individual domains which are interconnected by the rubbery segments B, and the structure imparts very high toughness to the material.

When the segments B dominate in the copolymer, the material is elastic. High elasticity is associated with the presence of large reversible deformations of polymers (for example during elongation), i.e. recovery to the original dimensions after removal of the deformation stress. The aggregate, immobile domains of segments in the glassy state prevent the irreversible translation of strained molecules in the block copolymer. The soft-segment chains are oriented in the direction of the deformation stress. The order of the system increases under stress and its entropy decreases. The entropy decrease is reflected in the retractive force oriented to the return of the sample into its original state.

As regards block copolymers, it is emphasised that only three- and multiblock copolymers show high elasticity. Two-block copolymers are unable to form a physically crosslinked structure since the soft segments only surround but not interconnect the glassy domains. The hard particles can therefore flow in the rubbery continuum of the two-block copolymer on deformation, and thus AB two block copolymers resemble in their properties unvulcanized filled rubber.

High elasticity is not only a property of some block copolymers but primarily it is a key feature of chemically-crosslinked amorphous linear macromolecules [22]. The reversible elastic property of crosslinked polymers is one of the few examples where the macroscopic behaviour of chemically modified polymers can be predicted from their structure. The molecular theory of rubber elasticity was derived for ideal polymer networks formed from chains following a Gaussian distribution of their end-to-end distances. Force f is connected with elongation on uniaxial deformation by the relation

$$f = (v_e RT / L_i)\eta(\alpha - \alpha^{-2})$$

where elongation $\alpha = L/L_i$ is defined as the ratio of the length of deformed and undeformed samples and v_e is the number of elastically effective chains in the network; the section of the macromolecule between two neighbouring crosslinking points is regarded as a chain. The parameter η is determined by the ratio $\langle r_i^2 \rangle / \langle r_o^2 \rangle$ of the mean square end-to-end distances in the undeformed network and in an identical assembly of free chains.

On deformation of the crosslinked polymer, the distribution of conformations is changed for each chain in the network. Hence, the entropy of the chains alters and the retractive force arises which is directed against the force of deformation and towards the state with the orginal equilibrium distribution of conformations. The stress in the deformed crosslinked polymer is to some extent an analogy of the pressure in a gas. Both phenomena depend on the kinetic energy of the elements of the system, i.e. on the molecules in the gas and on the chain segments in the networks, thermal motions of which result in the formation of the most probable coiled structure of the chain. In comparison with "normal" solids where only the energy of elasticity is present, the entropic character of elasticity is dominant in rubber-like polymers. The energy term, connected with the energy inequivalence of the various conformations of the chains, contributes to the network elasticity to a lesser degree.

The above theoretical equation describes the deformation behaviour of

elastomers undergoing small strains. The real elastic properties are better accounted for by the semi-empirical equation

$$f = (C_1 + C_2/a)(a + a^{-2})$$

where, in addition to the constant C_1 corresponding to the factors relating to the
theory of rubber elasticity, the constant C_2 describes the effects of intermolecular
forces and of chain entaglement.

The initial Young's modulus from the theory of rubber elasticity is given by

$$E = 3v_e RT\eta / V_i$$

where V_i is the volume of the undeformed network. It is seen from the equation
that the equilibrium modulus increases proportionately with temperature in
contrast to the uncrosslinked polymer. The modulus is also proportional to the
network density and can be utilized to determine the number of elastically
effective chains v_e. Assuming that all mers in the chains are involved in the
parameter v_e, the equation for the modulus can be rewritten as

$$E = 3\varrho RT\eta / M_c$$

where ϱ is the density of the sample and M_c is the number average molecular
mass of the network chains. In an effort to clarify the network topology, a
distinction is made between v_e, the overall elastically effective crosslinking including entanglements, loose chain ends, and similar "physical" effects, and v_c,
the number of real chains existing in the network (i.e. the degree of chemical
crosslinking). However, reliable estimation of the actual number of chains
formed in the crosslinking reaction is very difficult.

An appreciable *"shape memory"* is observed in those crosslinked polymers
where, during processing, the oriented molecules can fix their position by
physical crosslinking. After heating the sample above the temperature of formation of stable physical crosslinks, the sample changes its shape to that appropriate to chemical crosslinking only. For example, crosslinked polyethylene can
be partially formed above the melting temperature of its crystallites. After
cooling the material, the product retains its pre-formed shape. However, by
re-heating the sample above the melting temperature of the crystallites the
original shape of the products is recovered. The "shape memory" is utilized in
the production of shrink-wrap films and tubes (*Table 9.7*) and crosslinking
mainly increases the shrinkage stress. Some residual shape memory is also
exhibited by the uncrosslinked polymer because of chain entanglement.

Among the positive effects of crosslinking on mechanical performance, we
should mention the increased resistance to crack formation and improvement of
the shape stability in the polymer products. These properties and the possibility
of application at higher temperatures follow from the reduced flow of crosslin-

204

Type of film	Relative shrink stress	Strength/ MPa	Ductility/ %	Shrinkage/ %
crosslinked	20—40	80	150	70
original	1	20	700	55

ked macromolecules (relative to uncrosslinked) under static loading and dynamic strain.

Failure by break or other types of disintegration of the polymer product is not an instantaneous act but a kinetic process. The rate of deterioration of the mechanical strength of polymers depends on the applied stress. The disintegration of the polymer system is usually accompanied by *mechanochemical processes*. Even elongational flow of macromolecules is frequently connected with the disruption of chemical bonds in the chain backbone or of the intermacromolecular crosslinks. The resulting macroradicals initiate chain degradation and oxidation visualized by cracks in the stressed polymer material. The importance of mechanochemical processes in the deformation of polymers is supported by the fact that radicals are also formed in the processing of uncrosslinked polymers in the plastic (rubbery) state. Even under these conditions of relatively large mobility of the chains, their chemical structure is changed and, in general, they display a higher reactivity than undeformed macromolecules. The changes in polymers due to mechanochemical reactions on cyclic, repetitive deformation of elastic polymers are known as *material fatigue*. During each deformation cycle, a number of macromolecules are broken. The resulting macroradicals are annihilated by the inhibitors present and thus the degradation chain reaction cannot develop in the initial stage of usage of the material. However, degradation will continue after exhaustion of all compounds deactivating the mechanically-produced macroradicals.

The degradation induces deterioration of the mechanical properties of polymers through the activity of low-molecular-mass products and mainly as a result of the decrease in molecular mass. The presence of small molecules which reduces intermolecular cohesion and increases the chain mobility is manifested in a reduction in T_g and in the hardness of the polymeric system. The enhanced mobility may relax some internal stresses and temporarily increases the strength of the material. The situation is even more complex when, in addition, crosslinking reactions proceed, as in the case of high-impact polystyrene where, in the initial stage of photooxidation, the bending and elongation strength increase but the ductility and impact resistance decline rapidly (*Fig. 9.8*).

205

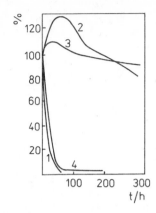

Fig. 9.8. Relative changes in the mechanical properties of tough polystyrene as a function of photo-irradiation time (xenon lamp). Original value — 100 %

(1) ductility, (2) bending strength, (3) elongation strength, (4) impact toughness

In the radiolysis of polymers where crosslinking dominates, the elongation, bending and shear strengths, impact resistance and elasticity modulus all increase in the initial stages; however, all these parameters diminish following irradiation to higher doses. The gaseous products formed on irradiation are partially retained inside the polymers and bring about additional stress, and the formation of cracks and other defects.

F. ELECTRICAL PROPERTIES

Polymers are often used as insulators in electrical devices because they behave as dielectrics in electric fields. The electroinsulation properties of polymers follow from the restricted mobility both of electrons and ions in the viscous media of nonpolar macromolecules.

Chemical reactions aimed at improving the electroinsulation are carried out chiefly with the object of prolonging the life-time of the dielectrics in an electric field and to increase the electric strength at elevated temperatures. Accordingly, the modification reaction needs to change the whole complex of material properties, which depend more on the assembly of chains than on the structure of individual macromolecules. Therefore, crosslinking or other reactions need to be considered which enhance the degree of crystallinity of the material or modify its crystal morphology. To improve the dielectric properties at the molecular level, the elimination of anomalous polar groups and hydrogenation of residual unsaturated bonds could be useful. Although these modification reactions have not yet been applied, they have potential in the tailoring of polymers to achieve materials with extreme electroinsulating and dielectric properties.

206

The electrical properties of polymers are subject to deterioration by oxidative degradation, which creates polar functional groups. Even if the relative permittivity does not change very much, the loss factor at some frequencies increases substantially and the polymer resistance decreases (*Table 9.8*). Under the action of oxygen and the electric field, the polar groups and low-molecular-mass products of degradation accumulate in the structural inhomogeneities of the polymer, diminish the electrical strength and finally result in the breakdown of the insulator.

Table 9.8 **Variation in the electrical properties of branched polyethylene induced by irradiation with a carbon arc in the presence of air (wavelength of irradiation 320—360 nm with λ_{max} 360 nm)**

Period of irradiation h	Relative permittivity at 5 MHz frequency	Loss coefficient at 5 MHz	Surface resistance $/10^8$ MΩ
0	2.28	0.02	210
16	2.27	0.05	200
23	2.25	0.06	65
48	2.28	0.11	49
97	2.29	0.23	39
129	2.30	0.28	25
155	2.31	0.27	5

The low conductivity of the majority of commercial polymers is a desirable property in most cases. Nevertheless sometimes it is responsible for a variety of drawbacks resulting from the build-up of static charge, for example as regards the danger of explosion in storage silos and tanks, interference in electrical circuits, the undesirable adhesion of packaging materials, the attraction of dust particles, etc. The half-life of the spontaneous discharge from the polymer into the environment varies from fractions of a second to several tens of hours; the rate of discharge increases with reduction of the surface resistance of the polymer. The electric charge on polymers is created through friction with other materials. Triboelectric charging and spontaneous discharge are surface phenomena and it suffices to link hydrophilic (sulphonate, carboxyl, hydroxyl, amino) groups to the polymer in a thin layer [23].

The electrical conductivity of some polymer system substantially increases on illumination. The increase in conductivity by about three orders of magnitude is ascribed to an increased concentration of charge carriers, mostly positive ion-radicals and cations (called holes), after electron transfer from the polymer, usually poly(vinyl carbazole) matrix to a suitable acceptor (e. g. 2,4,7-trinitro-9-fluorenone).

Chemical modification of the polymer matrix can extend the spectral region of the radiation which induces the photoconductivity, improves the sensitivity and optimises the mechanical properties of photoconductive macromolecular systems. Ionizing radiation temporarily increases the electrical conductivity of all polymers.

The interest in conducting polymers stems from the possibility of combining the unique properties of macromolecular systems with the conductivity of metals or solid-state semiconductors. In this connection, the elasticity, ductility, resistance to impact and chemical reagents, the possibility of orienting the structural units and of preparing materials with an anisotropic character, and ease of processing, are all properties particularly sought. In addition, there is the prospect of preparing, by the chemical transformation of polymers, electrically conducting elements spatially separated by the insulating regions of the original polymer. Delineation of the modified zones in the continuum of the original polymer can be accomplished for instance by locàl pyrolysis. The major problem of this approach is the dependence of the resulting conductivity of carbonised products on the pyrolysis, conditions and the rather low value of this conductivity. For example, the conductivity of the carbonised residue from the pyrolysis of polyacrylonitrile is from about 10^{-6} to 10^{-3} S cm^{-1}. The controlled elimination of side-groups from vinyl polymers, such as dehydrochlorination of poly(vinyl chloride), the dehydration of poly(vinyl alcohol), the elimination of acetic acid from poly(vinyl acetate), etc., yields unsatisfactory results. The resulting polyenes involve numerous defects and hence the conductivity of the polymeric products of such elimination reactions does not exceed 10^{-8} S cm^{-1}.

The method most widely used in the preparation of conducting polymers is the doping of polyconjugated macromolecules by an electron-acceptor or donor. The conductivity is determined both by the dopant concentration (*Fig. 9.9*) and the type of polymer and dopant (*Table 9.9*). A correlation evidently exists between electrical conductivity and the homogeneity of the polymer backbone (*Table 9.10*). It is not restricted to chains containing heteroatoms for reduced

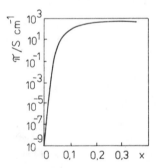

Fig. 9.9. Plot of electric conductivity of cis-polyacetylene against the molar ratio x of dopant to C=C units

Table 9.9   The specific conductivities of doped *cis*-polyacetylene and poly(1,4-phenylene)

Polyene	Dopant	$\dfrac{\text{mol dopant}}{\text{mol C=C}}$	S cm^{-1}
$\left[\text{CH=CH}\right]_n$	Na	0.42	25
	I	0.50	360
	IBr	0.30	400
	AsF$_5$	0.28	560
$\left[\text{—⬡—}\right]_n$	I	0.20	10^{-4}
	AsF$_5$	0.30	500
	Na	0.19	3 000

Table 9.10   The specific conductivity of some polyenes doped by arsenic (V) fluoride

Polyene	$\dfrac{\text{mol AsF}_5}{\text{mol C=C}}$	S cm^{-1}
$\left[\text{—⬡—O—}\right]_n$	0.13	10^{-3}
$\left[\text{⬡}\right]_n$	0.33	10^{-3}
$\left[\text{⬡—S—}\right]_n$	0.33	10^{-2}
$\left[\text{—⬡—S}\right]_n$	0.33	1
$\left[\text{—⬡—}\diagup\right]_n$	0.19	3
$\left[\text{CH=CH}\right]_n$	0.20	400
$\left[\text{CH=CH}\right]_n$	0.10	1 200
$\left[\text{—⬡—}\right]_n$	0.13	500

conductivity is observed even in copolymers of acetylene with 1,4-phenyl units. Heterogeneous centres cause the localization of charge and create an energy barrier to its movement [24].

The planarity of the polyenes is an important structural parameter in their conductivity [25]. However, planarity is not a necessary condition for conductivity, for example, the transfer of charge in nonplanar poly(4-phenylene sulphide) is mediated by the overlap of the *p*-orbitals of sulphur with π-orbitals of the phenyl rings.

The temperature dependence of the conductivity of doped polymers (*Fig. 9.10*) is less steep than indicated by the Arrhenius equation and follows the expression

$$\sigma = A \exp(-aT^{-0.5})$$

where σ is the conductivity, T the absolute temperature and A and a are constants. This less steep dependence of conductivity on temperature accords with a model of charge transport between the conducting regions and through the insulating barriers by means of a hopping or tunnel mechanism [26].

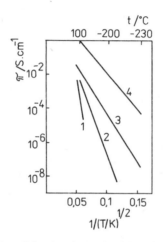

Fig. 9.10. Temperature variation of the electric conductivity of poly(1,4-phenylene sulphide) doped with AsF$_5$ (*1, 2, 3, 4* — increasing concentrations of dopant)

The electrical conductivity of polyacetylene doped with iodine depends considerably on its method of preparation, a difference by six orders of magnitude in conductivity being observed for equivalent concentrations of dopant. These differences correlates well with the degree of crystallinity of the samples; increase in the amorphous phase strongly diminishes the conductivity. Iodine is chemically bound to polyacetylene and in this way the conjugation of the polymer is disrupted. Since the addition reaction proceeds mainly in the disordered regions, the conductivity of polyacetylenes with a low degree of crystallinity is most affected.

The conductivity should also be influenced by the existence of branching and junction points in the networks, by impurities emanating from the synthesis, by the presence of oxygen and its overall time of contact (or reaction) with the polyacetylene. Drawbacks of polyacetylene and of many other polyenes are their insolubility and their decomposition below the melting temperature of the crystals.

Doped polyenes do not show classical ionic conductivity [27]. The long-term application of direct current induces no reduction in conductivity and no formation of charges in the vicinity of the connected electrodes. The number of charge carriers (holes) in poly(1,4-phenylene) doped with arsenic(V) fluoride is of the order of 10^{24} dm^{-3} and their mobility is about 10^{-4} m^2 V^{-1} s^{-1}. On doping with potassium, the average mobility of the electrons as charge carriers is lower, at a value of 10^{-6} m^2 V^{-1} s^{-1}.

References

1. Bikson, B., Jagur-Grodzinski, J., Vofsi, D.: Heterogeneous Reactions of Polymers: 4. Melting Behaviour of Heterogeneously Chlorinated and Sulphonated Polyethylenes. Polymer, *24*, 583—586, 1983.
2. Martinez-Salazar, J., Rueda, D. R., Cagiao, M. E., Lopez-Carcos, E., Calleja, F. J. B.: Influence of Chlorosulphonation on the Surface Mechanical Properties of Lamellar Polyethylene. Polymer Bulletin, *10*, 553—557, 1983.
3. Quirk, R. P.: Stereochemistry and Macromolecules. J. Chem. Educ., *58*, 540—543, 1981.
4. Briskman, B. A., Rozman, S. I.: Influence of Irradiation on Polymer Density (in Russian), Inzh. Fiz. Zhur., *43*, 288—293, 1982.
5. Brydson, J. A.: The Glass Transition, Melting Point and Structure, in: Polymer Science, Jenkins, A. D. (Editor), North Holland Publ. Co., 1972, Vol. 1, p. 194.
6. Peppas, N. A., Bussing, W. R.: Friedel-Crafts Crosslinking Methods for Polystyrene Modification: T_g Transitions of Grafted Polystyrene. Polymer, *24*, 898—902, 1983.
7. Meeten, G. H., Navard, P.: The Cholesteric Nature of Cellulose Triacetate Solutions. Polymer, *23*, 1727—1731, 1982.
8. Tverdochlebova, I. I.: The Basic Hydrodynamic Characteristics of Polymers (in Russian). Usp. Khim., *46*, 1280—1301, 1977.
9. Van Krevelen, D. W.: Properties of Polymers, Elsevier, Amsterdam 1972.
10. Roovers, J. E. L., Bywater, S.: Modification of the Solution Properties of Polystyrene in Cyclohexane by the Incorporation of Small Amounts of Foreign Groups. Macromolecules, *9*, 873—874, 1976.
11. Charlesby, A.: Atomic Radiation and Polymers. Pergamon Press, London 1960, *533*, p. 172.
12. Rejtlinger, S. A.: Permeability of Polymer Materials (in Russian). Chimija, Moscow 1974.
13. Rogers, C. E., Yamada, S., Ostler, M. I.: Modification of Polymer Membrane Permeability by Graft Copolymerization, in: Permeability of Plastic Films and Coatings to Gases, Vapors and Liquids. Hopfenberg, M. B. (Editor), Plenum Press, New York 1974, p. 155—166.
14. Leeper, H. M., Wright, R. M.: Elastomers in Medicine. Rubber Chem. Technol., *56*, 523—556, 1983.
15. Hamenesh, C. L., Dynes, P. J.: The Effect of Plasma Polymerization on the Wettability of Surfaces. Polymer Lett. Ed., *13*, 663—668, 1975.
    Sanžarovskij, A. T., Kovarskaja, L. B.: Adhesion of Polyethylene to Metals (in Russian). Dokl. Akad. Nauk SSSR, *203*, 409—412, 1972.

16. DECKER, Ch.: Surface Protection of Poly(Vinyl Chloride) by Photografting of Epoxy — Acrylate Coatings. J. Appl. Polym. Sci., *28*, 97—107, 1983.

17. REGEN, S. L., KIRSZENSZTEJN, P., SINGH, A.: Polymer — Supported Membranes. A New Approach for Modifying Polymer Surfaces. Macromolecules, *16*, 335—338, 1983.

18. RIGGS, W. M., DWIGHT, D. W.: Characterization of Fluoropolymer Surfaces. J. Electron Spectr. Rel. Phenomena, *5*, 447—460, 1974.

19. SCHONHORN, H., RYAN, F. W.: Surface Crosslinking of Polyethylene and Adhesive Joint Strength. J. Appl. Polym. Sci., *18*, 235—243, 1974.

20. KAWAGUCHI, H., HOSHINO, H., AMAGASA, H., OHTSUKA, Y.: Modification of a Polymer Latex. J. Coll. Interface Sci., *97*, 464—475, 1984.

21. AKLONIS, J. J.: Mechanical Properties of Polymers. J. Chem. Educ., *58*, 892—897, 1981.

22. MARK, J. E.: Rubber Elasticity. J. Chem. Educ., *58*, 898—903, 1981.

23. GIBSON, H. W.: Control of Electrical Properties of Polymers by Chemical Modification. Polymer, *25*, 3—27, 1984.

24. BAUGHMAN, R. H., BREDAS, J. L., CHANCE, R. R., ECKJARDT, H., ELSENBAUMER, R. L., IVORY, D., M., MILLER, G. G., PREIZIOSI, A. F., SHACKLETTE, L. W.: Macromolecular Metals and Semiconductors: A Comparative Study, in: Conductive Polymers Seymour, R. B. (Editor). Plenum Publ. Corp., New York 1981, p. 137—148.

25. TRIPATHY, S. K., KITCHEN, D., DRUY, M. A.: Molecular Structure Calculations in Two Classes of Conducting Polymers. Macromolecules, *16*, 190—192, 1983.

26. SHACKLETTE, L. W., ELSENBAUMER, R. L., CHANCE, R. R., ECKHARDT, H., FROMER, J. E., BAUGHMAN, R. H.: Conducting Complexes of Polyphenylene Sulphides. J. Chem. Phys., *75*, 1919—1927, 1981.

27. SHACKLETTE, L. W., CHANCE, R. R., IVORY, D. M., MILLER, G. G., BAUGHMAN, R. H.: Electrical and Optical Properties of Highly Conducting Charge — Transfer Complexes of Poly(*p*-phenylene). Synthetic Metals, *1*, 307—320, 1979.

X. CURRENT AND POTENTIAL APPLICATIONS OF THE CHEMICAL TRANSFORMATIONS OF POLYMERS

The properties of polymers are currently modified mostly via permanent changes in the chemical structure of macromolecules tailored for a particular application. In this way, starting from available macromolecular substances, the spectrum of materials required in applications related to construction and protection is further extended. Macromolecular reactions are also important as the unwanted side processes accompanying the technological isolation or processing of polymers. Future applications will call for chemical, electrochemical, mechanochemical and other reactions of polymers which mediate the exchange of information and energy with the environment. The enthusiasm and expectations aroused in this area are supported by the first successful steps already made, some of which will be mentioned in the following sections of this Chapter.

A. MODIFIED POLYMER MATERIALS

Interest in the modification of polymers in practice predates determination of the structure and macromolecular character of several natural materials. This was the case with the tanning of leather, and also much later in that of the chemically more advanced but still purely empirical investigations of the vulcanization of rubber (1838) and the nitration of cellulose (1845).

The main emphasis in the vulcanization of rubber was on making this seemingly very peculiar material harder and less sticky. Interestingly the Indians of Central and South America used rubber latex not only to make textiles and shoe materials water-resistant, but also to make bouncing balls, long before the era of Columbus. Chemists become more deeply interested in rubber three centuries later only when bicycles and cars arrived with their insatiable requirements for tyres. The nitration of cellulose was discovered some years later and all these events marked the beginning of major chemical influence upon everyday life. The chemical investigation of the reactions of cellulose was promoted by the fact that it is one of the most abundant macromolecular substances in the living world. It forms a basic component of plants, being found in cotton in a chemically pure form.

Cellulose is produced industrially from wood which serves as an example of an ideal polymer composite. Cellulose forms a fibrillar skeleton in wood which is immersed and fixed into noncellulosic polysaccharides (hemicelluloses) and the polyphenolic polymer, lignin. The fibrillar structure of cellulose accounts for the strength of wood, the hemicelluloses provide flexibility while the densely-crosslinked lignin gives it toughness. On the same principle, a combination of other properties can be achieved in various modifications of polymers and of multicomponent systems.

The process of cellulose production provides an example how decomposition reactions are utilized to isolate one component of a composite. Many valuable macromolecular substances are discarded in this technology, but nevertheless this method for the elimination of the noncellulosic components of wood was devised and has contributed heavily for years to river pollution. This situation results from imperfect technology, the relative accessibility of wood and from the importance of paper of which fibrillar cellulose is a basic component. Pressure for the production of pulp and cellulose also arises from the applicability of viscose rayon, cellophane films and other derivatives of cellulose. Their industrial preparation is based on the chemical transformation of cellulose molecules. At first, cellulose is converted to alkali cellulose by means of sodium hydroxide and then into cellulose xanthate by means of carbon disulphide. The highly viscous aqueous solution of cellulose xanthate is shaped into fibres or films in a precipitation bath consisting of a solution of sulphates where xanthate side-groups are eliminated and cellulose macromolecules are regenerated as the product of required shape.

In this short excursion into the history of a part of applied chemistry we have concentrated mainly on some successful developments based on the reactions of cellulose. It goes without saying, however, that this sector of polymer chemistry is not confined to history and is not limited to cellulose.

a. Polymers with Heteroatom Backbones

Cellulose as a multifunctional alcohol is relatively easily esterified by acids [1]. Of the numerous cellulose esters known, cellulose nitrate is produced in largest amount. Its properties and application depend on the degree of esterification (*Table 10.1*). Cellulose dinitrate plasticized with camphor is celluloid, which has played an important role in the past. (We may note that development of table-tennis was associated with the possibility of manufacturing the light, hard and elastic celluloid balls). Similarly the manufacture of photographic film based on cellulose nitrate enabled the expansion of the film industry. Nitro

N, % wt	Degree of substitution	Solvent	Application
11	2.33	ethanol	celluloid production
11.5	2.44	acetone	film-forming components of lacquers
12	2.55	esters	transparent and flexible
		ketones	supports of photographic emulsions[a]
13.5	2.86	esters	explosives
		ketones	

a) adhesives for paper, leather and woods; cements

lacquers as protectives of wood and metal have long been of great importance. The most serious disadvantage of the cellulose nitrates is their inflammability. Conversely, its reactivity and instability, and the large quantity of gases which comprise the reaction products of cellulose nitrate are the bases for its role in the manufacture of gun-cotton.

On esterification of cellulose by acetic anhydride, cellulose triacetate is obtained, and the less-substituted derivatives are prepared by its subsequent hydrolysis. The most reactive group (at C—6) reacts first during hydrolysis of the triacetate, and the resulting cellulose diacetate, with its accessible hydroxyl group is hydrophilic. In contrast, cellulose dinitrate is hydrophobic. This apparent paradox is a result of the different routes of preparation leading to the substitution of different hydroxyl groups.

The major advantage of the cellulose acetates is their self-extinguishing following ignition. The higher flame-resistance gave greater security to film-projectionists and the cinema audience and it also enabled the manufacture of acetate silk fibres. Owing to their reasonable thermal stability, cellulose acetates could be processed by methods typical for thermoplasts. However, they cannot compete with the commodity thermoplasts because of their more difficult processing and higher costs. As to the cost, one must note that, in order to modify a unit mass of cellulose, more than an equivalent amount of acetic anhydride must be used, which is more expensive than ethylene, styrene or vinyl chloride.

Besides cellulose acetates, the tripropionate is also manufactured and used for similar products as the triacetate in those cases when a greater water resistance is required at the expense of lower strength. Yet finer tailoring of properties can be achieved with products from cellulose acetate-co-butyrate, i.e. mixed esters of acetic and butyric acids in a ratio from 2:1 to 1:2 which amounts about 60 % of the total weight. Acetate-co-butyrates with a higher content of acetate groups are suitable for very tough films, and copolymers with

215

a majority of butyrate groups are used as the film-forming components of lacquers.

The second largest group of cellulose derivatives, the cellulose ethers, are prepared by reacting alkali cellulose with an alkylating agent. Using methyl chloride, methyl cellulose with a degree of substitution of between 1.5 and 2 is manufactured. This derivative, which is soluble in cold water, is used, inter alia, as a textile finishing material, a colloidal stabilizer of emulsions and latex paints, a thickener of food, a pharmaceutical and cosmetic product, a paper glue, a component of washing powders and for the conservation of offset plates in polygraphy.

Carboxymethyl and hydroxyethyl cellulose find the same type of application. To prepare the former derivative, the sodium salt of monochloroacetic acid is used as the esterification agent. The degree of esterification used in practice is between 0.5 and 1.2. Hydroxyethyl cellulose is produced by the addition of ethylene oxide to the hydroxyl groups of the anhydroglucopyranoside ring. Since polymerization competes with the simple addition reaction, the resulting products involve both the hydroxyethyl and poly(ethylene oxide) side-substituents. Accordingly this derivative could equally well be regarded as a grafted copolymer.

Ethyl cellulose with a degree of substitution between 2.3 and 2.6 is a water-insoluble derivative. It is used for similar purposes as cellulose esters soluble in organic solvents and especially in those situations where a resistance to hydrolysis in an acid medium is called for.

The cellulose derivatives and even cellulose itself are sometimes crosslinked subsequently. This procedure provides the anti-crease and anti-shrinkage finishing of textile fibres particularly from cotton and viscose. In this treatment, N-methylol derivatives of urea, which are more acceptable as regards hygiene than the volatile formaldehyde, are coated onto the fibre. Impregnated fibres are heated up to 130—160 °C in the presence of catalysts such as NH_4Cl, $ZnCl_2$, $ZnSO_4$, etc. The methylol derivatives condense with the hydroxyl groups of cellulose and form crosslinks between the macromolecules, restricting fibre shrinkage after swelling in warm water. The crosslinked fibres also display a higher strength when wet.

Several applications of originally water-soluble derivatives of cellulose are based on inhibition of their solubility or reduction of their swelling. For example, crosslinked carboxymethyl cellulose or sulphoethyl cellulose are used as cation-exchange resins, reversible adsorbents of water-soluble substances and dialysis membranes. The anion-exchanging cellulose contains aminoethyl or diethylaminoethyl groups.

In connection with water-soluble derivatives of cellulose, we might also mention an effort made several years ago to use them as a substitute for regenerated cellulose. Carbon disulphide, used in the manufacture of cellulose by the viscose process is very inconvenient because of its toxicity and potential for explosion. With hindsight we know that the intended substitute for the viscose process was unsuccessful mainly because crosslinked water-soluble derivatives exhibit mechanical properties inferior to those of regenerated cellulose.

As regards the remaining types of polysaccharide, crosslinking reactions are used mainly in starch. Crosslinked starch has not the consistency of a gel, but is a paste, i.e. the property often appreciated in food products.

Crosslinking is also used to modify the properties of proteins. The application of this reaction with the largest manufacturing output is certainly the chemical treatment of leather. The mechanism of crosslinking of proteins has been outlined in the Chapter V. It is unnecessary to enumerate the areas of application of this material attested by generations. New applications are restricted by the shortage of natural leather on the world-wide scale.

In order to improve certain properties of wool fibres, particularly reduction of their shrinkage on washing, the protein substrate is modified by bifunctional agents. The crosslinking of casein, isolated from milk, takes place in the processing of this product to give wool-like fibres. Casein, a copolymer consisting of 18 types of aminoacid residue, after swelling with water is firstly shaped and the product immersed in formaldehyde. In this medium, the protein chains are connected by crosslinks an the product hardens. The material is used mainly by craftsmen in the fabrication of ornaments such as buttons and brooches.

As regards synthetic polymers featuring heteroatoms in macromolecular backbone, the subsequent crosslinking is an unavoidable stage in the processing of polysiloxane rubbers. Crosslinked polysiloxane are relative costly but very valuable rubbers which are elastic in the temperature range from -100 to $250\,°C$. Besides their exceptional heat- and frost- resistance, they are non-toxic and possess good electroinsulation. Thus, they are ideal for exacting multipurpose applications such as in cable insulators, in aircraft and satellites, as a prosthetic devices in medicine, etc. At present about 300 000 surgical operations per annum are performed in which crosslinked polysiloxane rubber is implanted in the human body. The replacement of heart valves by silicon ball-valve implants amounts to 2 % of these operation.

b. Polymers with Carbon Backbones

Poly(vinyl alcohol) is a polymer which can be produced only by the substitution of side groups on a pre-existing macromolecule and not by polymerization of the corresponding monomer; it is prepared by hydrolysis of poly(vinyl acetate). Poly(vinyl alcohol) is soluble in water, which may need to be warm depending on the molecular mass and the degree of crystallinity of the sample. The number of residual acetate groups in the macromolecule affects its properties. Poly(vinyl alcohol) is used for the production of fibres, as a protective colloid in heterogeneous systems and as a raw material in the manufacture of poly(vinyl acetals). The latter are the products of condensation of two vicinal hydroxyl groups of poly(vinyl alcohol) with aliphatic aldehydes such as formaldehyde, acetyldehyde and butyraldehyde. Besides acetal groups, poly(vinyl acetals) contain residual hydroxyl and acetate groups.

Poly(vinyl formal) is employed in the manufacture of electroinsulating lacquers and adhesives. Lacquers and everyday consumer products are also manufactured from poly(vinyl ethanal). Films of poly(vinyl butyral) are employed in safety glass to prevent its breakage. This poly(vinyl acetal) has also proved of use in production of lacquers.

Polyethylene, the most widely employed thermoplastic material, is modified for industrial use by crosslinking, and to a lesser degree in the production of chlorosulphonated and chlorinated derivatives. Chlorosulphonation of polyethylene is a favoured method for the manufacture of fire-resistant rubber suitable for the insulation of cables in mines. Its resistance to ozone is another excellent property of this rubber which is utilized in electroinstallations of high voltage equipment. Chlorinated polyethylene is added to poly(vinyl chloride) to improve its toughness.

The crosslinking of polyethylene deserves more detailed discussion. The mutual linking of its chains considerably modifies the properties and areas of application of this polymer [2]. Generally, its mechanical properties are improved including the region above the melting temperature, and in this way, the temperature range for the application of polyethylene is broadened. An enhanced resistance in its response to an electric field while under mechanical strain is utilized in the high-voltage insulation of cables. The increase of viscosity and the reduction of its temperature dependence (*Fig. 10.1*) enable the manufacture

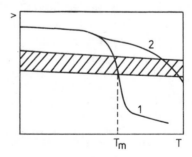

Fig. 10.1. Viscosity-temperature plot for molten uncrosslinked (*1*) and crosslinked (*2*) polyethylene. In the hatched region, the viscosity is optimal for formation of a foam; T_m — melting temperature of the crystallites

of light polyethylene foam with a density less than 20 kg m^{-3}. The blowing agents, which are finely distributed in the crosslinked polyethylene, form a homogeneous cellular structure of pores following the released gas which, in contrast to the uncrosslinked polymer, remain stable in this state without coalescence of the small bubbles into larger ones.

Polyethylene foam is used in the thermal insulation of buildings and vehicles, as shock-absorbing packing material, in life-saving waistcoats, etc. Crosslinking is important in the manufacture of thermally-sintered products with interconnected systems of pores; porous materials of this type are used as filtration media, in the controlled capillary transport of liquids as in the tips of colour pencil markers, in the spreading of functional liquids on moving objects, etc. Crosslinked polyethylene fibres can act as reinforcements in composite materials [4]. The crosslinking of films enhances their shape-memory, hence, after crosslinking the same shrinkage stress is achieved with material with one-half of the thickness of uncrosslinked film. This reduction in the required thickness of the films represents a direct material saving and the correspondingly reduces mass of polymer, brings about a shortening of the transit time in the heating tunnel of the packaging machine with a consequent energy saving and improvement in productivity.

The crosslinked structure also enables the addition of up to 60 % (wt) of inorganic fillers into polyethylene and the possibility of new applications especially in the construction industry.

An unexplored area in the crosslinking of polyethylene concerns the condensation reactions of brominated monocrystals. Apart from the formation of a network structure in the layers of the amorphous phase on the lamellar surface, the preparation of cyclic polyethylene appears possible, with rings of regular size, some of them being mutually interlocked to give crosslinking of the catenane type. The size of the resulting rings should be controlled by the conditions of crystallization of the original polyethylene.

Crosslinking has also found practical utilization in other polyalkenes. Crosslinked ethylene-propylene copolymer (1 : 1) is more resistant to oxidation during weathering than natural rubber, but its mechanical properties deteriorate. In addition, its adhesion to cellulose and polyamide fibres is low, which reduces its applications in tyre production. The copolymer of isobutylene with a small amount of isoprene (from 2 to 4 mol %) gives, after vulcanization with suphur, the so-called butyl rubber, with a low elasticity on rebound as utilised in shock absorbers. Its low gas permeability is the basis of its second principal area of application, namely the manufacture of inner tubes for tyres. The tubes from halogenated butyl rubber show even better maintenance of pressure over time. Butyl rubber with ethylene-propylene rubber amount to about 10% of the total consumption of synthetic rubbers.

Crosslinked polydienes represent the main category of synthetic rubber. This group involves copolymers of butadiene with styrene and acrylonitrile, 1,4-polybutadiene, cis-1,4-polyisoprene and polychloroprene. Butadiene copolymers, which emerged as the first substitutes for natural rubber, are the most widely-used types of rubber even today. In mixtures with other rubbers they form the principal component in the manufacture of tyres, conveyors, flexible junctions of pipes, packings and floorings. In the form of latex, they are constituents of adhesives and the film-forming components of protective and decorative paints. The products from pure cis-1,4-polybutadiene rubber release less heat on deformation than the products from natural rubber and display better wear resistance. As regards tyres, their lower adhesion to wet asphalt roads is their only disadvantage. It is curious that production of cis-1,4-polyisoprene, which accurately simulates the structure of natural rubber, amounts to only about 10% of the total output by weight of polydienes. This paradox results from the later discovery of stereospecific polymerization. The more expensive polychloroprene is produced in even lower amounts (about 5 %); its properties are well-balanced with good resistance to weathering and against oils.

Materials based on natural rubber or polydienes display varying degrees of hardness depending on the degree of crosslinking, i.e. mostly on the sulphur content during vulcanization. The widely-used soft rubber contains, apart from fillers, a maximum 10 % (wt) of sulphur, while the semi-hard and very hard rubbers contain up to 25 and 50 % of sulphur, respectively. The applications of rubber products are widely known and the main areas have been mentioned earlier. We should point out, however, that there are many other, unconventional applications such as foundations for earthquake-proof buildings, antivibratory blocks in the construction of bridges and underground transport systems, etc. One can speculate on some futuristic applications of crosslinked elastomers. Just as in model aeroplanes where the reversible deformation of rubber is used to drive a propeller, one can imagine the reverse process in the temporary conservation of the kinetic energy of a braking car and its restoration as the car accelerates.

Crosslinking is not the only way of modifying rubber. Some important applications are based on the cyclisation of natural rubber and polydienes. The cyclic rubbers are thermoplastic, hard materials used for adhesives, printing inks and as negative resists in microelectronics.

Chlorinated poly(vinyl chloride) has been produced for a long time but only on a small scale. Its first application was as the film-forming additive in paints and lacquers; later uses were as a moulding material and as a polymeric component in glues based on poly(vinyl chloride). Improvement of its mechanical properties in the temperature region 80—110 °C could be achieved but only with a high level of chlorination, when the material is hard to process due to its high melt viscosity. Hydrogen chloride is thermally eliminated much more slowly even with a small content of additionally bound chlorine, evidently due to the deactivation of the most reactive anomalous units in the poly(vinyl chloride) chain. The latter aspect of additional chlorinated material could be more significant than the improvement of mechanical properties.

c. Graft and Block Copolymers

In the early days of polymer science, all innovations, whether accomplished by systematic research or by chance, have found rapid application in industry. The processes of polymer grafting provide a rare exception to this rule. Grafting methods have been widely and thoroughly investigated and enable very useful combinations of the properties of polymeric materials, nevertheless only a few procedures have reached the stage of regular production and mostly in limited amounts. The reason underlying this strong interest combined with a limited application is the interconnection between the considerable variability in the properties of grafted polymers and the complications in their multi-step synthesis, thus the improvement in properties is not balanced by the higher associated costs in the case of the commodity polymers produced on the large scale.

One of the most important industrial applications of the concept of grafting combines the properties of elastic and hard polymers in the production of high-impact polystyrene, poly(vinyl chloride) or ABS (acrylonitrile-butadiene-styrene) copolymers. High-impact polystyrene is produced by the polymerization of styrene containing 5—10 % (wt) of natural rubber or another polydiene soluble in the polymerized monomer. Apart from diene polymers, "statistical" ethylene-propylene copolymer, ethylene-vinyl acetate copolymer and chlorinated polyethylene with about 25% (wt) of chlorine are used as the grafting components in the manufacture of high-impact poly(vinyl chloride). The applications of this type of grafted copolymer are similar to those of the parent thermoplastics but the products have greatly improved toughness (up to 5 to 10 times higher) and incomparably better impact resistance at low temperature. A small amount of elastomer mechanically mixed into the thermoplastic also improves its properties but the final effect is much lower. The high-impact thermoplastics are non-transparent two-phase materials since the dispersed particles of the grafted elastomer have a larger size than the wavelength of the incident light; any smaller particles of the rubbery phase would not improve the toughness of the material.

The properties of the elastomeric component predominate in grafted polymers with 40—60 % of vinyl monomer; the branches of the polyvinyl thermoplastic function as an active filler and physically crosslink the grafted copolymer. Accordingly these copolymers have the properties of vulcanizates filled with carbon black but are lightly coloured. Another advantage is that the improved flow of the molten copolymer in hot processing allows more precise moulding of the products. Grafted polymers containing a majority of elastomeric component are used to a lesser extent than high-impact copolymers but some areas of application can be mentioned. Natural rubber grafted with styrene is suitable for the manufacture of tennis balls. The grafting of methyl methacrylate onto poly-1,4-*cis*-isoprene gives a copolymer which is used in solution as an adhesive of the poly(vinyl chloride) films with rubber mouldings. The grafting of vinyl chloride onto the polydiene substrate produces a plasticized poly(vinyl chloride) without addition of a small-molecule plasticizer. This material is suitable for the packaging of food since food is not contaminated by the plasticizer even on direct contact.

Of block copolymers, that of ethylene and propylene is the commercially most successful. This material exhibits good mechanical resistance to long-term loading and a good impact strength at temperatures down to $-20\,°C$. Since these materials display shape stability at temperatures to $100\,°C$ and higher, the products can be sterilized and washed with hot water. The copolymer is suitable for production of reusable food containers. The handling of containers made with this copolymer is less noisy than those made of wood or aluminium.

Thermoplastic rubbers based on the three-block (ABA) copolymers, polystyrene-diene polymer (hydrogenated)polystyrene, find wide application. They have proved successful in the production of shoes and other products by injection moulding. They are used as effective moderators of the properties of polyethylene, polystyrene and polypropylene [5]. The processability of polyalkenes or polystyrene is improved by addition of 5—25 % (wt) of the block copolymer, bringing 20—40 % energy savings to the moulding process. In addition, incorporation of a three-block copolymer into polystyrene improves its impact resistance ten-fold. The toughness of polystyrene films is also enhanced by a two-block copolymer, which is better in this case than the three-block type as regards processability. Addition of block copolymers into polyethylene prevents, for example, breakage of sacks made from polyethylene. The shock resistance of polypropylene at temperatures below $0\,°C$ is improved by these copolymers. In another important modifying effect both of block and also graft copolymers, better miscibility of two homopolymers is achieved by the addition of copolymers comprising the same types of mer as the homopolymers. In this way a quasihomogeneous mixture resembling a stable emulsion is prepared rather than a true solution.

Block copolymers with both hydrophilic and hydrophobic chains have found various applications. They were firstly used as non-toxic surfactants in medicine and cosmetics (toothpaste, body lotion, cholesterol-reducing drugs, contrasting suspensions in diagnostic techniques, etc.) and are frequently applied as an antistatic additive in the washing of textiles.

B. SEMI-PERMEABLE MEMBRANES

Thin polymer membranes with a thickness of several μm are used as separation barriers in the concentration and purification of chemical compounds [7]. This particular application depends on the diameter of the channels and the type of porous matrix. Membranes of maximum pore size are prepared by the sintering of powdered polymers combined with the mechanical multidirectional stretching of semicrystalline films. Depending on the size of the powdered particles, the pore diameter lies between 0.1 μm and 50 μm.

Such membranes are particularly suitable for the filtration of suspensions and of polluted air. They also serve as supports for finer semi-permeable membranes. The cracks in mechanically-strained films are of dimensions between 20 nm and 50 μm. The mean dimensions of the cracks are governed by the degree of elongation of the film. Semi-permeable membranes of this type are used as battery separators and as a packaging material for one-time-use medical instruments, allowing for sterilization by gaseous ethylene oxide and preventing penetration by bacteria.

Microporous membranes with identical cylindrical pores are prepared by etching of irradiated (originally homogeneous) polycarbonate or polyester films. The polymer films with a thickness 10—20 μm are bombarded by a stream of charged reactor particles perpendicular to the film surface. The polymer is degraded locally along the track of the particles through the film and is removed selectively then by a suitable solvent. The irradiation time determines the pore density whereas the channel diameter is controlled by the extent of solvent extraction. Membranes are produced with pore diameters in the range 0.02—20 μm and are used for instance in the sterile filtration of biological solutions.

The most widely used microporous membranes are prepared from cellulose derivatives. In this technique the polymer is dissolved in a suitable solvent and the solution is cast on a support of thickness about 20—200 μm. Subsequently, the vapour of a poor (or non-) solvent is allowed to diffuse into the liquid film. The partially precipitated polymer forms a microporous structure with a pore size about 0.1—10 μm. Membranes of this type are suitable for the filtration of turbid solutions, for the separation of bacteria and some enzymes and for the detoxication of blood in artificial kidneys. Asymmetric membranes for ultrafiltration and reverse osmosis are prepared in a similar way using a faster feed of the non-solvent vapour.

Particles are transported through homogeneous membranes as a result of a pressure difference, a concentration gradient or an electric field. The separation of components depends on the rate of transport of the particles through the membrane. The transport rate is given by the diffusion coefficient and by the concentration partitioning of particles of a given type into the polymer film. This latter difference of the solubilities of the separated substances in the polymer matrix is an important characteristic of homogeneous membranes. By this mechanism an effective separation also takes place when the sizes of the particles are similar. The silicone film which separates helium from other gases is an example of a homogeneous membrane.

The thin homogeneous membranes (0.1—0.2 μm) used for ultrafiltration and reverse osmosis are combined with a highly porous membrane support with a pore diameter 100—200 μm. The support does not change the rate of filtration which is given by the thickness of the actual barrier layer. Thus the asymmetric membranes have a considerably better separation performance than symmetric membranes of an equivalent thickness. As a further advantage, the combined asymmetric membranes are not clogged by the separated particles since these remain on the surface of the barrier layer.

Ion-exchange membranes have positive or negative charges situated on the pore walls or inside the homogeneous gel. A membrane with attached positive ions is anion-exchanging and conversely for cation-exchange membranes. The operation of a charged membrane proceeds by the separation of counter-ions from dissolved salts and depends on the concentration of charges and ions in both the membrane and the solution. The charge density in good ion-exchange membranes is about 2—5 milliequivalent per gram of the membrane matrix. Ion-exchange membranes are prepared either by dispersal of an ion-exchanger into a polymer matrix or by the chemical incorporation of sulphonic acid, carboxyl or quaternary ammonium groups into the membrane skeleton. Ion-exchange membranes are applied chiefly to the electrodialysis of salt solutions in water.

The driving force of micro- and ultrafiltration and of reverse osmosis is the hydrostatic pressure difference on both sides of the membrane [8]. For microfiltration (filtration of particles of 1—10 μm), a pressure of from 10 to 200 kPa is used. A similar pressure is also necessary for ultrafiltration (separation of particles of about 0.3 μm). The separation of particles of different size proceeds by a sieve mechanism. When rather short macromolecules with a molecular mass about 3000 have to be separated, the osmotic pressure is so large that a pressure of about 2—10 MPa is necessary to overcome it. A symmetrical membrane is used in microfiltration whereas an asymmetrical one is required for the other two processes. In the ultrafiltration membrane the surface polymer is porous and its chemical identity does not influence the separation. In contrast,

224

the surface polymer film for reverse osmosis is homogeneous and the chemical structure of the polymer is of primary importance to its separation efficiency.

The driving force of dialysis derives from the concentration gradient of the solutes. The transport of small molecules by this mechanism is common in biological systems. Dialysis is used in the removal of low-molecular-mass compounds from a solution of macromolecules to a solvent separated by the semi-permeable membrane.

The electric potential difference is the driving force for transport in electrodialysis. In this process, ion-exchange membranes of opposite charge are placed between the anode and cathode in an alternating sequence. The aqueous solution of salts is pumped between the membranes and, under a direct current, the charged cations move to a cathode. The cations easily pass through the negatively charged cation-exchanging membranes but are held up at the positively charged anion-exchanging membranes. The final result of the selectively hindered movement of the ions through the membranes is an alternation in the concentrations of salts between the parallel membranes. The removal of sodium chloride and other salts by electrodialysis is the most economical method up to a concentration of 5 g dm^{-3} of salt in water. At higher concentrations, reverse osmosis becomes cheaper.

Membrane separation processes are of great importance in the food and pharmaceutical industries for the isolation of technical enzymes [9], for desalination of sea-water, purification of waste-water, for the production of high-purity water for the microelectronics industry, for the regeneration of reagents from dilute solutions, etc.

C. FUNCTIONALIZED POLYMERS

In principle one can select from two strategies in the functionalization of polymers. Functional groups and fragments of small molecules with the required specific action (drugs, pesticides, stabilizers, catalysts, dyes, etc.) are linked either to crosslinked or to soluble macromolecules. The attachment of functional groups to insoluble crosslinked organic macromolecular carriers is more widespread. Carriers of the "gel" type are based on slightly-crosslinked networks where the degree of swelling varies with the diluent used and the pores are of small diameter. In contrast, the heterogeneously-crosslinked "macroreticular" carriers involve regions in which the aggregate structure of the chains is not perturbed by the diluents. Large pore dimensions are characteristic of the latter type of carrier but their total surface area is normally lower than for the gel carriers.

In selecting a suitable carrier for attachment of a particular functional group, one has to optimize several contradictory requirements. Apart from the effective surface area and pore size, one needs to take into account the distribution and shape of the pores, the mechanical strength and the volume changes of the gel particles induced by the solvent, the distribution of grain size and the polarity of the gel. Polystyrene, polyacrylamide, poly(glycol methacrylate), modified cellulose and dextran are the polymers most widely used in the preparation of crosslinked carriers. In conclusion, we should mention that functional groups can also be linked to inorganic materials such as silica gel and porous glass.

a. Polymer Reagents

Functionalized polymers, either soluble or linked to networks, have numerous advantages over conventional reagents. They can be easily separated by filtration, centrifugation and selective precipitation, and permit the use of a large excess of reagent in order to enhance the rate or yield of reaction. The excess reagent can be easily purified and recycled by the batch method or in a column. The problems of lability, toxicity, volatility and odour are much reduced when working with polymer reagents. In addition, the possibility exists of designing the particular microenvironment of the reaction centre corresponding to the requirements of the reaction as regards steric effects, polarity, etc.

The polymer reagent may serve as a transfer agent for functional groups in its reaction with low-molecular-mass reagents. For example, polystyrene with bound triphenylphosphine dichloride transforms carboxylic acids into their chlorides, alcohols into chloroalkanes, ketones into chloroalkenes, etc.

The application of polymer-bound reagents incurs some disadvantages, for instance, a lower reaction rate than the corresponding reaction of small molecules in solution, and greater difficulty in monitoring the course of reaction due to the presence of the polymer chain. Nevertheless their numerous merits guarantee their rapidly increasing utilization in the near future.

Polymer-bound reagents have played a seminal role in protein synthesis by the method known as *the Merrifield synthesis* [10]. The procedure provides for the routine synthesis of proteins of a specified primary structure with an almost unlimited number of amino acid residues in the chain. Since at least two reactive groups, amine and carboxyl, are in each amino acid (A), one of the groups has to be blocked. The synthesis proceeds step by step

$$Y - NHA^1CO_2H + X - \textcircled{P}$$
$$Y - NHA^1CO_2 - \textcircled{P}$$
$$Y - NHA^2CO_2H + NH_2A^1CO_2 - \textcircled{P}$$
$$Y - NHA^2CO_2 - NHA^1CO_2 - \textcircled{P}$$
$$\vdots$$

peptide fragment $+ \textcircled{P}$

226

In the first step, amino acid A^1 is bound to the polymer carrier via its carboxyl group. The blocking group Y is removed from the now-attached fragment and the second amino acid is linked to the growing chain. Finally, at the end of the synthesis, the polypeptide chain is detached from the polymer carrier. Analogous methods are now available for the syntheses of nucleic acids and polysaccharides. Repetition of the numerous relatively simple steps enables automatization of the whole process of assembly of biopolymers.

b. Polymer Catalysts

Catalysts attached to crosslinked polymers represent an intermediate level between homogeneous and heterogeneous catalysis. Currently they can replace the original, homogeneously dispersed catalysts even on the industrial scale. The advantages of both homogeneous catalysis (selectivity) and heterogeneous catalysis (the facile separation and recycling of the catalyst) are retained. This type of process has, apart from its technical advantages, numerous ecological benefits both directly through reduction in output wastes and indirectly in the reduced degree of corrosion, and consumption of energy and raw materials. Depending on the character of the catalyst, it can be attached to the carrier either directly by a covalent bond or in some cases simply by physical adsorption, complexation or by entrapment in the polymer matrix.

Ion-exchange resins were used in catalysis well before than the more complex, ingeniously functionalized polymer catalysts become available. For instance, cumene is produced from a mixture of propylene and benzene in the presence of H_2SO_4, the amount of sulphuric acid being as much as 150 kg per ton of product. The replacement of liquid acid by a "solid" version, e. g. by sulphonated poly(styrene-divinylbenzene) resin could eliminate the problems associated with the use of sulphuric acid [11].

The functional groups on polymers and copolymers are suitable sites for the immobilization of complexes of transition metals. Sometimes the valence state of the metal is stabilized by its attachment to the polymer chain and thus the reduction of the activity of catalyst with time is retarded.

When a catalyst is insoluble in the reaction medium it is sufficient to cover the surface of a carrier by a catalyst such as $TiCl_3$ or VCl_3. For a soluble catalyst such as $(C_2H_5)_2TiCl_2$ or $VO(OR)_3$ a chemical bond has to be formed which survives the catalytic process, as in the carrier based on allyl alcohol

$$(-CH_2-CH-)_n + n\ VO(OR)_3 \longrightarrow (-CH_2-CH-)_n$$
$$\underset{\underset{OH}{|}}{CH_2} \qquad\qquad \underset{\underset{\underset{O=V-(OR)_2}{|}}{O}}{CH_2}$$

In this way, incorporation of the entire organometallic unit of the transition metal can be accomplished, as for example in the one-component catalyst for ethylene polymerization

$$(-CH_2-CH)_n + n\, Ti(C_7H_7)_4 \longrightarrow (-CH_2-CH-)_n$$
$$\qquad\quad |\qquad\qquad\qquad\qquad\qquad\qquad\quad |$$
$$\qquad\quad CH_2 \qquad\qquad\qquad\qquad\qquad\qquad CH_2$$
$$\qquad\quad NH_2 \qquad\qquad\qquad\qquad\qquad\qquad NH$$
$$\qquad\qquad\qquad\qquad\qquad\qquad\qquad\qquad\quad |$$
$$\qquad\qquad\qquad\qquad\qquad\qquad\qquad\qquad Ti(C_7H_7)_3$$

The polyalkene support is modified at the surface in such cases in order to incorporate the functional groups allowing the immobilization of the catalyst. The catalyst is bound chiefly in the pores of the macroreticular carrier; in contrast to the usual small-molecule catalysts, the situation in the pores favours the formation of an effective catalytic centre because of the cooperative interaction between the individually bonded complexes [12].

Copper complexes are widely used as catalysts of oxidation-reduction reactions. From macromolecular chemistry we can cite the example of the synthesis of poly(2,6-dimethylphenylene oxide)

The oxidative polycondensation of 2,6-dimethylphenol proceeds in the presence of coordination complexes of Cu(II) with tertiary amines. The pyridine units linked to a crosslinked carrier may function as complexing ligands. Tight linking of the complex to the crosslinked support may influence the catalytic activity adversely, demonstrated by the molecular mass M of the resulting poly(phenylene oxide) [13]. When the copolymer of styrene, divinylbenzene and 4-vinylpyridine is used as the catalyst, the molecular mass of the poly(phenylene oxide) is about 14 000 due to steric hindrance by the polymer network in the coordination sphere of Cu^{2+} ions in the later stages of reaction. However, when using the polymer support where pyridine (Py) side-groups are linked to the network via a flexible spacer

$$Ⓟ—phenyl—CH_2OCH_2CH_2CH_2Py…Cu(II)$$

poly(phenylene oxide) with M about 24 000 is formed. This example documents the change in catalytic activity with variation in the conformational freedom of subunits in the catalytic centre. The attachment of the catalytic centre to a chain and subsequent modification of the functional groups in its vicinity enables the

228

design of an optimal catalytic environment similar to the situation found in enzymes. The active centre can be linked either to a soluble polymer as poly(ethylene imine) or to a crosslinked polymer.

The most direct way of technically exploiting a macromolecular system with the catalytic activity is perhaps through the linkage of an enzymatically active protein to a polymer support. Obviously, the attachment has to be realized via those sites which are not directly involved in the catalytic act. Using specially designed techniques, the enzymes are linked to the support by the end groups of the protein, by lysine side-groups, etc. The preservation of the biological activity of the enzyme is the criterion by which to judge the final success of an immobilization. The hydrophilic polymers such as agarose, cellulose, and cross-linked dextran are the most suitable carriers for the immobilization of enzymes. A sufficiently high degree of crosslinking makes the molecules inside the gel behave as in a viscous solution.

Apart from covalent binding, the enzyme can be attached to the polymer support by non-covalent adsorption. This interaction is particularly strong in the so-called affinity linkage based on the structural similarity of an enzyme substrate and of the polymeric support as, for example, in the case of α-amylase bound to cellulose. The disadvantage of this process lies in the possibility of dissociation of the affinity complex by variations in temperature, ionic strength and pH. The entrapment of enzymes in gels (mainly of the polyacrylamide type) is also frequently used; the enzyme is entangled into the network prepared in solution from monomers and a crosslinking agent.

In many applications of biocatalysis it is unnecessary to use purified enzymes. The immobilization of entire microbial cells is a less costly variant. This method is used in the fermentation production of penicillins, ethanol, citric acid and of some amino acids. The entrapment of whole cells in the gel, or their encapsulation into polymer particles, are apparently the most suitable approaches in this case. Optimization of the porosity, (the main requirement of the polymer network embedding the biological material), inhibits the escape of cells from the gel and yet allows fast transport of reactants and products and also of oxygen and nutrients.

c. Selective Sorbents

Functionalized and crosslinked polymers are used for the separation and isolation of a single or several, mostly minority, components from complex mixtures. In some cases, molecules or metal ions are adsorbed irreversibily onto the polymer which is then decomposed and the enriched component is isolated. A more convenient and desirable procedure is that in which the adsorbed component is liberated non-destructively from the polymer sorbent.

$$\text{(P)} - X + A, B, C, D... \rightarrow \text{(P)} - X...A + B, C, D \xrightarrow[\text{filter}]{\text{column}}$$

$$\text{(P)} - X...A \rightarrow \text{(P)} - X + A$$

Due to its simplicity and the effective regeneration, this approach has found wide applications in biochemistry, hydrometallurgy and in analytical chemistry.

In biochemistry, highly specific separation of biological macromolecules is accomplished by this method. For example, an enzyme can be isolated by its attachment in a column to a polymer bound to a specific inhibitor with an affinity for the enzyme. By the elution of a mixture containing the enzyme through the column, the enzyme-stationary phase complex is formed. The pure enzyme is then desorbed by the changing the conditions of elution. This technique, termed *affinity chromatography*, is also an effective means of isolating other types of biological complex such as antigen-antibody. It can also be used in reverse, i.e. the attached enzyme functions in the purification of inhibitors (*Fig. 10.2*)

Fig. 10.2. The concept of affinity chromatography. The enzyme E, attached to the carrier in the column, binds the inhibitor (or vice versa)

In hydrometallurgy, the essential operation is the extraction of metals from dilute aqueous solution by ligands. The latter can be linked to a polymer, for example, mercury ions can be bound to polymers with thiol groups and copper, nickel and other metals to poly(4-vinyl pyridine). Polymers with two or more ligand binding sites produce the complexes more readily, for example poly(4-vinyl-2,2'-bipyridine)

$$(-CH_2-CH-)_n$$

forms a strong complex with Cu^{2+} ions.

Polymers involving macroheterocycles of the crown ether type not only selectively bind the ions of alkali metals and alkaline earths but take up the ions of transition metals and lanthanides, organic ions and even some nonpolar compounds. The selectivity as regards the size and electronic structure of a cation for one type of polymer is illustrated by the 10^6 times stronger binding of Cd^{2+} cation than of the similar Zn^{2+} cation. Polymers with attached macroheterocycles have found applications in extraction, separation, enrichment and recovery of metals and radioisotopes, in ion-activated catalysis, as the active component of ion-selective electrodes, etc. The formation of complexes with some cations facilitates the solubility of some salts in nonpolar media.

Functionalized polymers and sorbents are increasingly important in environmental technology. Macromolecular polystyrene adsorbents are used in the purification of waste water, for the removal of phenol and other noxious compounds in the treatment of drinking water, and for the elimination of toxins from blood. Some other types of adsorbents have proved successful in the elimination of oil pollutants following disasters at sea and of pesticides from water. Certain nitrogen-containing polymer adsorbents, for example polymers with *N*-glycidylpiperazine groups, can selectively chemisorb SO_2 from air.

The mechanism of separation on polymer-based adsorbents is identical with those encountered in various other types of chromatography. In general, the separation takes place because of non-identical interactions of individual particles in the mixture with the stationary phase. A gas or liquid can serve as the mobile phase and, according to the prevailing separation mechanism, chromatography can be termed as adsorption, partition, steric exclusion, etc. The rate of separation and the resolution of the components is determined by the selection of the chromatographic carriers, mainly by their polarity, porosity, specific surface area, degree of swelling, etc. Developments in the chemical transformation of polymers targeted on the preparation of these new chromatographic carriers and their characterization will add considerably to the advancement of chromatography.

d. Macromolecular Analogues of Chemical and Biological Reagents

The chemical transformations of molecular fragments with a particular form of the reactivity usually proceeds by the same mechanism regardless of whether they are in low-molecular-mass compounds or in macromolecules. The attachment of a reactive functional group to a polymer usually brings about longer reaction times. The slower translational diffusion of macromolecules reduces the migration of functional fragments to and from the system targeted. This new property of "sustained release" is utilized in polymer-linked analogues of stabilizers, plasticizers, fire-retardants, drugs, pesticides and other compounds with a specific activity. The reduced migration of various additives from polymeric materials is also desirable as regards hygiene regulations. Environmental health considerations are significant in the case of the fire retardants added of necessity to some polymer products in high concentrations and which contain the elements of chlorine, bromine and phosphorus. Thus for example, the application of chlorophosphazene as a fire-retardant of cellulose textiles, is improved by covalent linking to the cellulose, a procedure which also reduces its losses in washing and dry-cleaning. Another possibility might be linkage of chlorophosphazene to chitosan, the natural linear aminopolysaccharide, because of its greater reactivity. With respect to the structural similarity of these two polysaccharides, chitosane adheres firmly to cellulose fibre after application from solution.

The agricultural protective compounds such as pesticides, herbicides, and fungicides are another group of compounds suitable for attachment to polymer chains. The active part can thus be liberated by gradual hydrolysis of the spacer group between the pesticide and a macromolecular chain. Enzymatic degradation of the spacer provides the most suitable avenue and thus cellulose derivatives such as carboxymethylcellulose are used as carriers. A copolymer which can be mentioned as an example of a pesticide with an extended period of activity due to its gradual elimination from the chain, is that formed by acetalization of poly(vinyl alcohol) by 2,6-dichlorobenzaldehyde [14].

The pesticide exhibits strong herbicidal and medium fungicidal activity and is eliminated from the polymer even with small concentrations of water. The retention of materials is also utilized in the application of polymers as carriers of perfumes, deodorants and pheromones as used in forestry and agriculture, etc.

A prolonged period of action is the primary goal in the functionalization of polymers with drugs. However, the process of attachment also influences the permeability of the drugs and their absorption by the individual types of cells, enabling, in principle, regulation of their selectivity. The polymer carrier, either soluble or insoluble, must first of all be unobjectionable hygienically and its linkage to biologically active compounds must lead to a well-defined derivative. Sometimes degradation of the carrier in the organism to produce non-toxic materials is desirable. In this case, the carriers of natural origin such as polysaccharides, or biological polyesters such as polyhydroxybutyrate or protein-related carriers are the most appropriate. The carriers are usually pharmacologically inactive but some polymers with biological effects have been tested. For example, polyethyleneimine might have some inhibitory effect on tumour growth and a strong immunostimulating effect has been found in the copolymer divinyl ether-maleic anhydride

The biologically active fragment is spontaneously eliminated from the carrier by hydrolysis, usually of the ester or ether link. Elimination of the drug from an inert carrier sometimes requires enzymatic attack and so the spacer needs to be designed with surroundings similar to those of the cleaved (ester) bond of the corresponding substrates. For hydrolysis by chymotrypsin, a phenylalanine group should be introduced

A variety of drugs have been linked to soluble carriers. Antibiotics such as tetracycline, ampiciline, streptomycine and others have been linked to the polysaccharide carriers, dextran and inulin. The resulting powdered water-soluble preparation show good antimicrobial properties. A molecular mass of the carriers of about 5000 was found to be optimal since then the carriers are safely excreted through the kidneys. The fixation of cancerostatic drugs is another attractive example. The antitumour drug methotrexate was fixed to a chain by a polymer-analogous reaction with divinylether-maleic anhydride co-polymer and enhancement of its activity was observed [15]. The cytostatic concentration was increased selectively in the tumour as a result of the attachment, whereas the drug concentration in other cells remained low. Another anti-tumour drug, cyclophosphamide, was anchored both to the previously mentioned carriers polyethyleneimine and divinylether-maleic anhydride copolymer and to poly(2-hydroxypropyl methacrylamide)

$$-[CH_2\overset{\displaystyle CH_3}{\underset{\displaystyle \underset{\displaystyle \underset{\displaystyle \underset{\displaystyle CH_2CH}{NH}}{CO}}{C}}}-]_n \quad OH \quad CH_3$$

This soluble carrier enables attachment of various fragments via transformation of the secondary hydroxyl group. Cyclophosphamide either alone, or linked to a carrier, is inactive against tumour cells in vitro but is activated in vivo after subsequent enzymatic oxidation [15]. Sometimes the anchored fragment does not function as a drug, but as a diagnostic tool of various processes in an organism such as the transport of substances, metabolism, etc.

In some cases, the biological activity cannot be ascribed to a single functional group but is determined by the presence of several functional groups or chain fragments. This may be illustrated by the effect of heparin on blood. Heparin is a mucopolysaccharide polyelectrolyte having three kinds of ionizable groups in the molecule. It is largely used as an antithrombotic agent preventing blood coagulation. This property proves to be connected with the presence of N-sulphonate and carboxyl groups in its chain, since a similar effect was observed with a cis-1,4-polyisoprene derivative of the structure

$$-CH_2\overset{\displaystyle CH_3}{\underset{\displaystyle \underset{\displaystyle SO_3Na}{NH}}{C}}----\overset{}{\underset{\displaystyle COONa}{CH}}CH_2-$$

Enzymes attached to hydrophilic gels by covalent bonds are used as diagnostics in medicine. Of interest also is the effort to immobilize protein hormones such as insulin at the gel without loss of their biological activity.

D. CHEMICAL TRANSFORMATIONS USED TO INVESTIGATE THE STRUCTURE AND DYNAMICS OF MACROMOLECULES

Designed reactions which attach, remove or block some easily detectable functional group in the polymer chain, represent a useful tool in improving our understanding of macromolecules. The specific activity of groups giving rise to fluorescence or spin interactions of the attached radical site is sensitive to changes

234

in the character of the polymer matrix. This may assist in the evaluation of the mobility of a macromolecular chain in close proximity to an attached "probe". We have already demonstrated (chapter II) the use of the photochromic spirobenzpyran group in visualizing the T_g transition of a matrix, which was based upon the isomerization reaction leading to decoloration of the sample. The spin labels involve paramagnetic groups such as in nitroxyl radicals derived from piperidine or pyrrolline

$$\begin{array}{c} O \\ \parallel \\ C-R^1 \end{array}$$

Through the reactive group R^1, the radical may be easily attached to a macromolecule. The shape of the corresponding ESR spectrum, which depends on the rate of spin rotation, reflects the local environment around the spin label. The changes in the spectra yield information on the local density of the matrix as well as on the intensity of segmental motions in the region of the attached group. Thus it was established that the rotational mobility of crosslinked polystyrene swollen with a good solvent is about 10 times lower than that in solution. The quantitative determination of the dependence of segmental mobility upon network density and upon different types of crosslink would certainly by of interest.

Fluorescent labels are frequently used in the investigation of the structure of biopolymers. The intense fluorescence of chromophores such as N,N-dimethyl-5-amino-1-naphthyl sulphonyl chloride is illustrative.

The shape of the fluorescence spectrum indicates the polarity in the microregion of the attached group while depolarization measurements yield information on the rotational mobility of the macromolecule, e.g. of proteins in a biological membrane.

A fluorescent label on a polymer may be also used to track the fate of a polymer introduced into a living organism, following the precedent of ^{14}C-labelled materials.

The chemical transformation of polymers has found application in diffraction or scattering methods, which are indispensable in the determination of macromolecular structure in the solid state. The problems here arise from the fact that the commonest polymers consist of atoms which diffract or scatter the incident radiation weakly. Useful progress in the X-ray structural analysis of monocrystals of globular proteins was therefore achieved only after the introduction of such heavy atoms such as Ag or Hg which replaced hydrogen atoms in the —SH groups of the protein molecules and thus enhanced the diffraction pattern. Since the original and chemically-modified proteins should give isomorphous crystals, this *method of isomorphous substitution* enabled determination of the spatial structures of myoglobin and haemoglobin.

Also in the case of synthetic polymers, it can become necessary to introduce atoms with large scattering factors. The scattering of X-radiation at low angles enables an estimate of the sizes of labelled macromolecules distributed in a matrix of unlabelled molecules. The dimensions of molecules of statistically brominated or iodinated polystyrene dispersed in unlabelled amorphous polystyrene were determined in this way. An unequivocal confirmation of the fact that the dimensions of macromolecules in the amorphous state and in dilute solutions under the theta condition are practically the same was achieved by means of low-angle neutron scattering. To induce the necessary contrast between modified and unmodified molecules, hydrogen atoms were substituted by deuterium. Such a substitution represents only a small perturbation in the chemical and thermodynamic properties and any objection as to whether labelling has any influence on the original arrangement of the macromolecules, may thus be discounted. Such isotopic substitution is used very frequently in vibrational spectroscopy and in the mechanistic study of chemical reactions. The peptide bonds of proteins are deuterated in order to obtain a distinct spectrum.

Along with the total substitution of some functional groups, deliberate *blocking of certain groups* may also be of value. Such an approach becomes important particularly when investigating the role of some groups in a protein molecule with respect to the net stability of its spatial structure and to the catalytic efficiency or specificity of enzymes. Modification of only one amino acid moiety may lead to significant changes of protein conformation; sometimes the spatial structure and active site of the protein remain unchanged and yet, the enzymatic activity disappears. Modification of lysozyme with iodine [16], for example, totally suppresses its enzymatic activity. X-ray structural analysis of deactivated lysozyme shows no appreciable changes of its molecular structure as compared with the unmodified enzyme. The only change due to the modification is that in the position of the Glu-35 unit and the formation of the internal ester bond with the Trp-108 unit. Taking into account the role of Glu-35 in lysozyme (see chapter II), the loss of catalytic activity of the modified reaction site is readily understood.

The mutual arrangement of structural units in a polymer chain may be deduced from the reactions of particular functional groups. This may be illustrated by the mechanism of dechlorination of poly(vinyl chloride) which has been determined with progressively greater precision. It was found [17] that the treatment of poly(vinyl chloride), dissolved in boiling dioxan, with zinc powder removed 84—87 % of chlorine from the polymer. Approximately the same percentage of bromine was eliminated from poly(vinyl bromide). The ozonolysis of halogenated polymer did not give uniform, low-molecular-mass products, as would be expected when only C=C bonds are formed on dehalogenation. It was assumed that the regular head-to-tail structural units

$$—CH_2CHCl(CH_2CHCl)_nCH_2CHCl—$$

in the polymer chain best accommodate the formation of cyclopropyl rings. A similar pattern of reaction was observed for the dehalogenation reactions of low-molecular-mass organic 1,3-dihalides. The upper limit of dechlorination as determined analytically accords with the idea [18] that

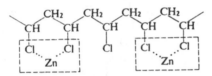

the randomly-formed, isolated CHCl groups are incapable of further reaction. Assuming a random initiation of infinitely long macromolecules, the calculated value of the maximum conversion on dehalogenation is 86.5 %. Such very good agreement between the experimental and theoretical values for the maximum degree of dechlorination was so convincing that the mechanism of dehalogenation of poly(vinyl chloride) was believed to have been determined unequivocally. Experiments with vinyl chloride copolymers [19], however, have revealed that the rate of dechlorination varies from copolymer to copolymer and that the upper conversion limit for dechlorination may be even higher for copolymers than for poly(vinyl chloride) alone. To explain this discrepancy, it was suggested that larger than 3-membered rings can also result from dechlorination, the process being accompanied by dehydrochlorination. As is evident from the formation of an insoluble fraction of dechlorinated poly(vinyl chloride) in more concentrated solutions, it may be also assumed that higher levels of dechlorination follow from the random reaction of reaction sites taking place on a single macromolecule as well as among different macromolecules.

Further insight into dechlorination comes from the observation of increased formation of double bonds in poly(vinyl chloride) following its reaction with zinc powder [20]. The unsaturation could result from either dehydrochlorination of thermally unstable structural units or dechlorination of anomalous head-to-head linkages in the polymer chain

$$—CH_2CHClCHClCH_2—$$

A new aspect concerning the mechanism of dechlorination was brought about by analysis of the NMR spectra of dechlorinated poly(vinyl chloride) which confirmed the formation of cyclopropyl rings on the one hand, but also revealed the appearence of new methylene groups in the chain of the dechlorinated macromolecule. This observation was interpreted in terms of successive hydrogenation of $C{=}C$ double bonds by hydrogen, possibly produced in the reaction of HCl with Zn. The radical centres at carbon atoms, which may alternatively be formed in the primary stage of the reaction with Zn, do not combine to form an aliphatic ring but rather abstract hydrogen from the surrounding solvent molecules to form CH_2 instead of the original CHCl groups.

Summarising, studies of the dechlorination of poly(vinyl chloride) satisfactorily deal with which kind of addition reaction prevails in the polymerization of vinyl chloride. In contrast to earlier opinions it has been shown that not only are cyclopropyl rings formed on dechlorination but also a relatively large fraction of the chlorine atoms is substituted by hydrogen. The mechanism of reaction seems thus to be justified but not yet proved. Unambigous experimental evidence as to the structure of the anomalous units in the poly(vinyl chloride) chain as well as their quantitative determination remains to be secured.

Chemical reactions are also utilized *in studies of* the *morphology* of polymer systems [22]. Since the amorphous zones of a semicrystalline polymer are usually more reactive than the crystalline regions, analysis of the reaction products gives some information on the regularity and conformation of the polymer chains.

Oxidation of polyethylene monocrystals by nitric acid leads, for example, to the formation of dicarboxylic aliphatic acids which have a relatively narrow distribution of molecular masses. Since the molecular mass of the resulting acids is considerably lower than that of the parent polymer, the macromolecules in a monocrystal are likely to be folded regularly. Nitric acid selectively attacks the folds at the monocrystal surface. The samples crystallized from a polyethylene melt and those additionally oriented give different gel permeation chromatograms after oxidation with nitric acid, which is an indication of different arrangements of the macromolecules in these samples.

The supermolecular structure of solid polymers may also be investigated by electron microscopy, where the mutual configuration of the structural units of macromolecules is made more distinct by the surface reaction of the reactive elements of a given heterogeneous system. Unfortunately the pictures obtained, which are sometimes very spectacular, also depend on the internal structure of the reacting surface of the polymer sample as well as on the mode and conditions of a subsequent 'etching' reaction [23]. Interpretation of the results is, therefore, often ambiguous even though the observed changes in the surface structure may be of use.

These various examples illustrate different possible applications of the deliberate transformations of macromolecules in the study of their structure and properties. The methodical approach indicated above will be applicable in the near future, particularly in the investigation of the complex functions of biopolymers under physiological conditions. It may also find use in the detailed investigation of the reactions of mutagenic and carcinogenic reagents with nucleic acids, in the clarification of the effects of enzymological repair at the molecular level and also in the study of the immunological reactions of polysaccharides and glycoproteins and in genetic manipulations.

E. MODELLING OF BIOPOLYMER FUNCTIONS

The existence of living matter is always associated with macromolecular systems where life functions represent a mutually conditioned sequence of chemical reactions and complex-forming equilibria of biomacromolecules. The characterization of these processes is normally the subject of biochemistry, enzymology, molecular biology and genetics as well of macromolecular chemistry. The interpretation of the mechanism of processes at the molecular level taking place in living systems includes modelling of such important processes as the recording of genetic information, the functioning of the brain, energy-transformation in muscles, visual process, and the immunological reactions of the organism.

a. The Energy-transformation Processes

Various transformations of one kind of energy into another may be realized through the chemical reactions of macromolecules. Such energy transducers are not currently of practical use but recognition of the rules governing them may lead to large-scale applications in the future. Biological systems which attain high efficiencies in the conversion of one kind of energy into another usually serve as examples in the postulation of model systems. Modelling of the direct *conversion of chemical into mechanical energy*, as it occurs in muscles has been widely quoted. In this case the transformation of energy is enabled by conformational changes in a polymer chain. Depending on the primary structure and surrounding medium, a macromolecule may adopt a particular secondary structure which determines its overall shape and dimensions. One extreme conformation may be represented by the rod-like helix shape, another by the random coil of a macromolecular chain. Changing the shape of the macromolecule changes concomitantly the average end-to-end distance and the supermolecular dimensions of the polymer sample. The extent of these changes depends on the regularity of the arrangement of the system as well as on the degree of chemical transformation of the primary structure of the polymer chains.

Some mechanisms of transformation of chemical and mechanical energy may be demonstrated by the change in dimensions of the partially oriented and crosslinked macromolecules of polyacids induced by changes in pH of the reaction medium (*Fig. 10.3*). Ionization of the functional groups of the polyacid brings about a repulsion of like charges which leads to stretching of the macromolecule and to lengthening of the 'synthetic muscle'. On reducing the level of ionization, the macromolecules recoil into denser coils and the fibres of the polyacid macromolecule contract. The helix-coil transformation and its associated changes of dimensions are demonstrated by collagen macromolecules.

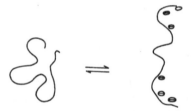

Fig. 10.3. Illustration of the stretching and contraction of the macromolecules in synthetic muscle due to pH changes of the medium

The ribbon constructed from oriented and partially crosslinked helical macromolecules becomes shorter following immersion into concentrated aqueous solutions of some salts such as LiBr. The helices become denatured by the salt to form macromolecular coils. On removing the salt or decreasing the concentration of Li cations, the helical structure of the collagen is restored and the ribbon reverts to its original dimensions, i.e., it becomes longer. The return to the original state is enabled by the partial crosslinking of the macromolecular chains which determine the overall orientation of a macromolecule in the sample.

Macromolecular systems can also transform *thermal energy* into *mechanical work*. This occurs as a consequence of the contraction of randomly coiled macromolecular chains in the elastic state by increasing the temperature of the polymer. In a non-isothermal heat engine, the amount of energy transformed depends on the temperature increase and on the overall change of entropy. With regard to the thermal stabilities and glass transition temperatures of polymers,

240

a heat engine working with polymers may withstand only a small temperature difference. They may, however, be used for the transformation of residual heat of some sources of relatively low temperature into mechanical energy.

Transducers of chemical and mechanical energy are capable of lifting masses a thousand times their own mass, which constitutes an immense output compared with heat engines based on gas or with electric motors. The fact that synthetic muscles have not found application in practice is due to their lower rate of energy transformation. To achieve more efficient energy conversion, systems with coupled enzymatic reactions should be designed.

The transformation of visible or ultraviolet light into chemical and mechanical energy represents another interesting possibility. In such a case, the macromolecules should contain photoactive chromophores. Changes in hydrophobic interactions may be observed in the photoisomerization of macromolecules with azobenzene side-groups (*Fig. 10.4*). The more stable *trans*-isomers which dominate in the dark are more easily aggregated and, due to hydrophobic interactions, they promote a tighter conformation of the macromolecule. After irradiation, the number of *cis*-isomers of the azobenzene side-group, which self-aggregate only slightly, increases and the polymer chain becomes more bulky. However, the efficiency of the energy transformation is once again low. Low efficiency is also characteristic of the transformation of light energy into electrical energy in organic polymeric photoconductors.

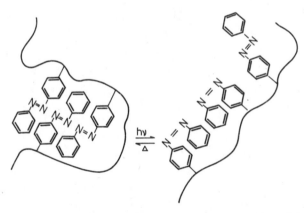

Fig. 10.4. Changes in the interactions of the phenyl rings and in stretching of the polymer chain after photoisomerization of the pendant azobenzene groups (Δ — the reverse reaction is induced by heat)

Enormous effort has been devoted to the transformation of solar radiation into chemical energy. The systems based upon methyl viologen (1,1′-dialkyl-4,4′-bipyridinium cation) ruthenium(II) complexes, which are able to initiate the photolysis of water into hydrogen and oxygen, seem most promising. The transport of electrons needs to be controlled to avoid side-reactions. Polymer carriers [24] provide an effective separation of the electrical charges as in biological redox systems of chlorophyll and cytochromes. Future technical achievements are likely to be based on the principle of cascade changes of by-products and excited states; one-step energy transducers may serve only as a first approximation to our understanding their functions. Photochemical isomerization of low-molecular-mass compounds for example may enable the collection and storage of light and its conversion into chemical energy. In some further step, the isomerized compound may be transported to a heat exchanger or reactor where an enzymatic catalyst will regenerate the parent compound and release the heat.

The reverse process, i.e. the transformation of mechanical energy into heat, has not been seriously considered in technical application. On the contrary, it is associated with problems such as the warming up of tyres on repeated deformation; of interest could be cooling processes based on macromolecular energy transducers.

Other kinds of energy may also be transformed by macromolecules [25]. Of particular interest is that of light into electrical energy in polymeric photoconductors such as poly(N-vinylcarbazole) doped with 2,4,7-trinitrofluorene but here again the efficiency of the process is lower than in the widely-used sil on solar batteries.

b. The Storage and Exchange of Information

Genetic information is transmitted from generation to generation via macromolecular nucleic acids utilising the sequence of four different structural units. Information on the structure, shape and function of a protein is contained in the sequence of about 20 different aminoacids of the polypeptide chain. In principle therefore it is possible to effect the miniature inscription of a text, determined only by a chain sequence of amino acids or other chemically different mers. Technically, this may be performed even today, along with the reverse deciphering of such recorded information. Considerable problems, however, arise from the time-consuming and very expensive operations associated with the recording and play-back. Moreover, during play-back the record will be destroyed. The practical applications of recording at the molecular level unavoidably require the evolution of new principles for the processing of stored information by non-destructive means based probably on the transcription of the chemical code into electrical or light impulses. The results obtained during the study of changes of information content in the primary structure of biopolymers as implemented in the functions of living organisms seem to be closer to practical use. These vital processes may serve as examples as to how the principles of transcription and exchange of information may be derived at the macromolecular level. As we know from molecular biology, DNA retains its genetic information for the next generation of some organism and transmits it by replication to the descendants. At replication, the macromolecular structure is copied by a positive-negative mechanism. The pair of complementary macromolecules in the helix is firstly unwound. Then, the synthesis of complementary macromolecules takes place on both filaments of DNA which finally leads to the two genetically identical double helices.

The parent information may be disrupted if replication competes with some other chemical reaction. Such an effect may be displayed by mutagens or carcinogens which react with the side-groups of the nitrogen bases of DNA. Radiation which initiates the crosslinking of two adjacent bases may also exhibit a mutagenic effect.

Hydrolytic reactions of DNA are brought about by restriction enzymes which specifically cleave the macromolecular chain into large fragments. Restriction enzymes are used to sequence nucleotides in nucleic acids and in genetic manipulation. The reverse, enzymatic combination of fragments of DNA originating from two different parent molecules incorporates different genomes into the macromolecule.

A more complicated transcription of genetic information from nucleic acids occurs during the biosynthesis of proteins. The sequence of structural units in DNA controls the protein synthesis; a particular sequence of three nucleotides corresponds to a particular amino acid in the protein chain.

In addition to the three-letter code, the positive-negative principle is effected in protein biosynthesis, especially in the gradual deciphering of the three-letter words in the DNA chain. The storage and the transfer of information via nucleic acids is associated with their ability to form complexes stabilized by hydrogen bonds and by the stacking interactions of the hydrophobic layers formed by the aromatic rings of the nitrogen bases. As is evident from antibody-antigen and drug-receptor interactions or from the specific enzymatic hydrolysis of proteins at certain amino acid groups, complexation equilibria involving macromolecules are of immense importance in living organisms.

Hydrolysis enables the reconstruction of protein aggregates and activates different protein precursors to reactive molecules. It involves the transformation of a proenzyme to the corresponding enzyme, such as those of trypsinogen to trypsin and a small peptide, the formation of hormones involving the transformation of proinsulin to insulin and the transformation of procollagen to collagen. Only in some cases, as for example those of the enzymes trypsin and chymotrypsin, the total proteolysis of the protein chain to individual amino acids occurs. These enzymes cleave the peptide bonds regardless of the structure of the vicinal amino acids. There are, however, various enzymes which only attack specific sites in the protein and thus decipher part of the information contained in its primary structure. Although many uncertainties exist about the detailed interactions of macromolecules, and will continue to exist for some time, the general picture of the receiving and transfer of molecular information and the chemical code for the transcription from nucleic acids functioning in protein synthesis seems to be understood. The practical applications of such knowledge for the modelling of memory and in operating systems awaits further research.

Another principle of information recording can be achieved by the designed distribution of macromolecules in some spatial unit. The differential solubilities between linear and crosslinked polymer or between parent and degraded polymer, both achieved after the effect of light, electron beams, etc., may be utilised. The information record at the supermolecular level may thus be compared with routine photography. The difference consists, however, in the sensitivity of both procedures. The high sensitivity of common photographic materials is based upon the fact that the microcrystal of AgCl in gelatine layer containing about 10^9 Ag atoms needs to be hit by only 5 light quanta to experience an appropriate effect. From such a microcrystal, all the silver ions are reduced by a developer; on the other hand, during the photocrosslinking of copolymers of cinnamic acid, one light quantum induces the insolubility o: 10^5 —10^6 atoms in the polymer chains of the photosensitive layer. From a comparison of the number of atoms in one recording element, the possibility of miniaturization leading to an increase in the information content by the use of photosensitive polymeric layers may well be effected. At the same time, the smaller the dimensions of the fundamental pixel of the picture, the higher is the capacity of the record, the unit area or volume then containing more pixels. Similar conclusions can be reached for the reverse process of polymer degradation.

The information recorded in the polymer layer may be used directly in polygraphy as offset prints or as the isolating or protecting shields in the production of printed circuit boards. The recording of light information mediated by a polymer may be further processed either by etching, by a further layer of semiconductor or by the various procedures of planar technology used in microelectronics.

Systematic investigation of the chemical reactions of polymers was instigated as a result of empirical improvements in the properties of natural products. The achievement of irreversible changes in the chemical structure of macromolecules became, in turn, the driving force for the designed modification of the properties of polymers. Modification of the chemical structure of polymers broadens the variety of materials available from existing polymers. A clear understanding of the reactions of macromolecules also serves in the optimization of the technological processing of polymers. Quite new horizons are opened by research on biochemical macromolecular systems and models. Information from these areas of application is already comprehensive although it has yet to go beyond the laboratory scale.

This limitation should not be a disincentive but rather a spur, since it is clear that we are about to witness the birth of a new generation both of macromolecules and of technologies appropriate to their exploitation. Finally, it is unlikely that the study of the transformation reactions of polymers will cease as both stable and activated macromolecules are the very substances of life itself.

References

1. BLAŽEJ, A., ŠUTÝ, L., KOŠÍK, M., GALIS, E.: Chemistry of Wood, (in Slovak), Alfa, Bratislava 1975.
   MORITA, M., SAKATA, I.: Chemical Conversion of Wood to Thermoplastic Material, J. Appl. Polymm Sci., *31* 831—840, 1986.
2. DORN, M.: Fortschritte auf dem Gebiet der PE-Vernetzung mit Organischen Peroxiden. Gummi Asbest Kunst., *35*, 608—611, 1982.
3. SAGANE, N., HARAYAMA, H.: Plastic Foam—Radiation Crosslinked Polyethylene Foam. Radiat. Phys. Chem., *18*, 99—108, 1981.
4. de BOER, J., PENNINGS, A. J.: Crosslinking of Ultra-High Strength Polyethylene Fibres by means of Gamma Radiation, Polymer Bull., *5*, 317—324, 1981.
5. HSIEH, H. L. BURR, R. H.: Block Polymers as High Efficiency Property Modifiers. Modern Plastics Internat., June 1982, p. 46—48.
6. SCHMOLKA, I. R.: A Review of Block Polymer Surfactants. J. Amer. Oil Chem. Soc., *54*, 110—116, 1977.
7. STRATHMANN, H.: Membrane Separation Processes. J. Membrane Sci., *9*, 121—189, 1981.
   MOLAU, G. E.: Heterogeneous Ion Exchange Membranes. J. Membrane Sci., *8*, 309—330, 1981.
8. KAMIDE, K., MANABE, S.: Mechanism of Permselectivity of Porous Polymeric Membranes in Ultrafiltration Process. Polymer J., *13*, 459—479, 1981.
9. PAŠEK, A.: Isolation of Technical Enzymes by Ultrafiltration. (in Czech), Chem. Listy, *75*, 856—869, 1981.
10. STEWART, G. M.: Solid Phase Peptide Synthesis. J. Macromol. Sci., Chem. A *10*, 259—288, 1976.
11. KLEIN, J.: Polymer Catalysts. A Contribution to Improve the Ecology of Chemical Processes. Macromol. Chem. Suppl., *5*, 155—178, 1981.
12. DYACHKOVSKII, F. S., POMOGAILO, A. D.: Synthesis and Catalytic Properties of Transition Metal Complexes Immobilized on Macromolecular Supports in Polymerization Processes. J. Polym. Sci., *68*, 97—108, 1980.
13. TSUCHIDA, E., NISHIDE, H.: Catalytic Behaviour of the Cu Complex Attached to Polystyrene Resin with a Spacer Group, Preprints Symp. on Macromol. Florence 1980, Vol. 4, p. 147—150.
14. SCHACHT, E., DESMARETS, G., GOETHALS, E., St. PIERRE, T.: Synthesis and Hydrolysis of

Poly(vinyl acetals) Derived from Poly(vinyl Alcohol) and 2,6-Dichlorobenzaldehyde. Macromolecules, *15*, 291—296, 1983.

15. PRZYBYLSKI, M., RINGSDORF, H., FUNG, W. P., ZAHARKO, D. S.: Synthesis, Antitumor Activity and Pharmacological Properties of Methotrexate Derivates Linked to the Divinylether — Maleic Anhydride Copolymer. Symp. on Macromolecules, Florence 1980, Preprints 4, p. 54—57.

16. SEGAWA, S., NAKAYAMA, M.: Difference in Substrate Binding Processes between Intact and Iodide—Inactivated Lysozyme. Biopolymers, *18*, 1503—1514, 1979.

17. MARVEL, C. S., SAMPLE, J. H., ROY, M. F.: The Structure of Vinyl Polymers. VI. Poly(Vinyl Halides). J. Amer. Chem. Soc., *61*, 3241—3244, 1939.

18. FLORY, P. J.: Intramolecular Reactions between Neighbouring Substituents of Vinyl Polymers., J. Amer. Chem. Soc., *61*, 1518—1521, 1939.

19. ALFREY, T., HAAS, H. C., LEVIS, Ch. W.: The Dehalogenation Reactions J. Amer. Chem. Soc., *73*, 2851—2853, 1951; J. Amer. Chem. Soc., *74*, 2025—2027, 1952.

20. SMETS, G.: Analogous Polymer Reactions, their Kinetics and Organic Chemistry. Pure Appl. Chem., *12*, 211—226, 1966.

21. TEPELEKIAN, M., PHAN WUANG THO, GUYOT, A.: Dechlorination of Poly(Vinyl Chloride) by Zinc. Europ. Polym. J., 795—805, 1969; CAIS, R. C., SPENCER, C. P.: The Dechlorination of Poly(Vinyl Chloride) by Zinc and Tributyl Tin Hydride. Europ. Polym., J., *18*, 189—198, 1982.

22. WEEKS, E., MORI, S., PORTER, R. S.: The Morphology of Ultradrawn Polyethylene I. Nitric Acid Etching Plus Gel Permeation Chromatography, J. Polym. Sci., *13*, 2031—2048, 1975.

23. NEPPERT, B., HEISE, B., KILIAN, H. G.: Ion Etching of Polymers. Colloid Polym. Sci., *261*, 577—584, 1983.

24. AGEISHI, K., ENDO, T., OKAWARA, M.: Electron Transport across Polymeric Membranes Containing the Viologen Structure. Macromolecules, *16*, 884—887, 1983.

25. ISE, N., AKIMOTO, T., TABUSHI, J., TAMASU, A., IMOSAEE, S., IKEHARA, S.: Speciality Polymers (Translated into Russian), Mir, Moscow 1983, Iwanami Shoten, Tokyo 1980.

INDEX